Why We Believe

FOUNDATIONAL QUESTIONS IN SCIENCE

At its deepest level, science becomes nearly indistinguishable from philosophy. The most fundamental scientific questions address the ultimate nature of the world. Foundational Questions in Science, jointly published by Templeton Press and Yale University Press, invites prominent scientists to ask these questions, describe our current best approaches to the answers, and tell us where such answers may lead: the new realities they point to and the further questions they compel us to ask. Intended for interested lay readers, students, and young scientists, these short volumes show how science approaches the mysteries of the world around us and offer readers a chance to explore the implications at the profoundest and most exciting levels.

Why We Believe

*Evolution and the Human
Way of Being*

Agustín Fuentes

Yale UNIVERSITY PRESS
NEW HAVEN AND LONDON

Templeton Press

Yale University Press books may be purchased in quantity for educational, business, or promotional use. For information, please e-mail sales.press@yale.edu (U.S. office) or sales@yaleup.co.uk (U.K. office).

Designed and set in Hoefler Text by Gopa & Ted2, Inc.

Printed in the United States of America.

Library of Congress Control Number: 2019937669
ISBN 978-0-300-24399-4 (hardcover : alk. paper)

A catalogue record for this book is available from the British Library.

This paper meets the requirements of ANSI/NISO Z39.48-1992 (Permanence of Paper).

10 9 8 7 6 5 4 3 2 1

To Humanity,
we who are neither accident nor miracle

Contents

Preface

BELIEF IS the most prominent, promising, and dangerous capacity that humanity has evolved.

Belief is the ability to draw on our range of cognitive and social resources, our histories and experiences, and combine them with our imagination. It is the power to think beyond what is here and now and develop mental representations in order to see and feel and know something— an idea, a vision, a necessity, a possibility, a truth—that is not immediately present to the senses, and then to invest, wholly and authentically, in that "something" so that it becomes one's reality.

Beliefs and belief systems permeate human neurobiologies, bodies, and ecologies, acting as dynamic agents in evolutionary processes. The human capacity for belief, the specifics of belief, and our diverse belief systems structure and shape our daily lives, our societies, and the world around us. We are human, therefore we believe.

SHERRINGTON'S CHALLENGE

Eighty years ago, when giving his famous Gifford Lectures in Edinburgh, the Nobel Prize–winning neurophysiologist

Charles Scott Sherrington mused[1] that the development of the human, in body and mind, was neither "accident nor miracle," that "organisms must be the sum of their parts and more," and that the mind "makes an effective contribution to life." What continued to elude the science of his time, he went on, was an explanation for how all of that could be true. While a complete and final answer to "the human" still eludes us (and maybe always will), we are far better able than Sherrington was in 1937 to offer insights about how humans evolved and how we develop in body and in mind.

In this book I take up Sherrington's challenge and update his questions:

▸ How do we understand humanity as neither the result of random processes nor the product of divine intervention?
▸ How can we be made up entirely of biological parts and organic processes and still dream, hope, and believe?
▸ How can our minds and our beliefs shape ourselves, other life around us, and even the planet itself?

If we can answer these questions, not just of human development, but of human becoming, of human believing, then we can step closer to the goals Sherrington sought.

Today we have a much better scientific understanding of the processes of human development and evolution than Sherrington did. Developmental biology, genomics, and evolutionary science have made enormous leaps in the past century and especially in the past few decades. The same is true for the study of the human past. Paleoanthropology, archaeology, anthropology, and neurobiology have given us a radically new landscape of understanding, of knowing, and of forecasting about ourselves, other life, and the whole planet.

In the first two decades of the twenty-first century we've redefined the very foundations of our evolutionary history, how our biology functions, and what it means to ask how we become who we are. Humans are neither accident nor miracle, and the explanation of who and why we are is an amazingly complex, dynamic, enticing, and unfinished story. It is a story in which belief is central, as both an outcome and a cause.

My own background is as an anthropologist, meaning I am trained in the biological and behavioral study of humans and our closest relatives. I have spent the past thirty years in deep engagement with the bodies, actions, and ecologies of humans past and present, of primates across the globe, and as an active participant in the debates about, and modeling of, evolutionary processes. I have long been enmeshed in enriching, enlightening, and maddening collaborations with a diverse array of scientists, philosophers, theologians, and other scholars. It is this type of transdisciplinary engagement, the cross-fertilization of ideas, methods, and theoretical grounds that I bring to bear on the data from human bodies and behavior past and present.

As an evolutionary scientist I try to uncover the specific origins, functions, and processes that undergird our capacity for belief. As a social scientist I seek to understand these findings in the context of the human experience: our social structures, belief systems, and daily lives. My goal in pursuing both of these pathways is to develop a better understanding of what it means to be human—past, present, and future.

In this book I share with you a story of our evolution rooted in the scientific endeavor, in the facts of our bodies, genes, ecologies, histories, and behaviors, but one that tries not to lose sight of the equally relevant philosophical narratives that run alongside and through the science. Unlike most evolutionary narratives, the one I present ties the explanation of humanity to our distinctive capacity for belief.

A Little Clarification of Terminology and Content

In this book I use the term "belief" to mean more than its basic definition of "trust, faith, or confidence in someone or something."[2] Belief is also a richer concept than the slightly antagonistic Wikipedia definition: "the state of mind in which a person thinks something to be the case, with or without there being empirical evidence to prove that something is the case with factual certainty."[3] It is not about being fooled.

The literary theorist Terry Eagleton, drawing on the philosopher Kierkegaard, tells us that the act of *Believing* is an act of being wholly and completely in love with a concept, an experience, a knowledge.[4] But believing is also an avenue to imagining and becoming, in ways that need not be rooted in the daily material reality but that can be infused with hope. Believing can be fearing an unknown but wholly felt entity or perception, or it can be a certainty of something that cannot be seen, grasped, or measured. Believing is completely real but often without material substance. And most critically, undergirding and infusing belief is the human capacity to imagine, to be creative, to hope and dream, and to infuse the world with meaning.

Belief, for better and worse, is a deeply and distinctly human process.

When hearing the word "belief," most assume it refers to some form of religion. Let me be absolutely clear: the human capacity for belief is not only about religion, spirituality, ritual, or some notion of the supernatural. It is not only our ability to have faith in something or someone, or our capacity for self-deception (even though these are important parts of the human experience). Throughout this book, I separate the *having of faith*—the specific content of belief—from the *capacity to have faith*, which arises from our core ability to believe. This is a critical distinction, especially when it comes to the belief systems that most would call "religion."

That said, religious belief is a major element in the human story and directly related to our capacity to believe, and thus one of the arenas on which we focus here. Yet I do not seek to explain or provide evidence for the emergence of any given set of faith practices. This is not something a scientist can honestly or effectively do, and anyway it is not what this book is about. I touch on specific beliefs, but only in passing, because the particulars of any given faith are best elaborated by the faithful. They cannot be explained by evolutionary scientists.

FOUR KEY QUESTIONS

Humans can see the world around them, imagine how it might be different, and translate those imaginings into reality . . . or at least try to. Meaning, imagination, and hope, which constitute our capacity for belief, are as central to the human story as bones, genes, and ecologies. As a species, we are distinguished by our extraordinary capacity for creative cooperation, our ability to dream big and make those dreams materialize, and our powerful aptitudes for compassion and cruelty—all of which are constructive of, and mediated by, our capacity for and practices of belief.

I propose that we can best understand *why we believe* by answering four key questions:

1. How do humans relate to the rest of the work biologically and ecologically?
2. What key evolutionary events and processes make us human?
3. How did the changes we made in the world enable the infrastructure for contemporary belief?
4. *How* do we believe? (What are the processes by which we believe?)

Adding these four together, we can offer an evolutionary answer to the question "Why do we believe?" As you will see, the answer allows us to examine specific categories of belief such as religion, economies, and love, and leads us to a final consideration: today in the twenty-first century, does belief still matter?

PART I

*Who Are We and How Did
We Come to Believe?*

CHAPTER 1

Belief, Evolution, and Our Place in the World

H ERE ARE THREE facts about who we are biologically:
   1. Humans represent an infinitesimally small percentage of all the life on this planet.
   2. Humans are deeply and substantially linked to all other life.
   3. Despite being a tiny part of the great diversity of living things, humans are among the most significant forces affecting all other life.

How we became so significant is one of the most important questions facing humanity. Our capacity for belief is a major part of the answer.

First, a little context.

Scientists have catalogued more than 2 million species of life, and they estimate that this is only about 25 percent of the species out there (most iving things are really quite small). While life is amazingly diverse, not all lineages are equally represented in the panoply of organisms. For example, the 400,000 or so species of beetles represent close to 20 percent of all named species. When the renowned evolutionary biologist J. B. S. Haldane was asked by a group of theologians what one could conclude about the nature of the Creator from a study of his creation, he is said to have answered, "He had an inordinate fondness for beetles."

The theological significance of beetles notwithstanding, we can unequivocally state that we humans are not very prominent in the overall picture of the diversity of life—at least on the face of things. Among the lineages of all living organisms, we don't really stand out. We are one of many, many lines of backboned organisms (a group called "chordata"), lumped in with monkeys, dogs, platypuses, iguanas, salmon, and chickens. Nevertheless, despite our anonymity in the panoply of life, we can learn from our lineage quite a bit about who we are, and how our ancestry lays a baseline for our capacity to believe.

The first time I looked at a cell through a microscope I was amazed. I remember thinking, *This is what we are made of. Each of these contains the secret of who we are. Buried deep in the nucleus is the DNA, the blueprint for life.* But my thought was far too simplistic. Like most people, I had a fantastical view of DNA, thinking it contains the code for who we are, or the instructions for making an organism. The truth is it has neither of these things. DNA is part of an amazing, intricate system of interrelated proteins, enzymes, and other molecules and chemical relationships that interact to enable core aspects of the development of organisms and their patterns of life.[1] DNA cannot do anything alone, and it does not contain either the secret of life or a blueprint. It does offer us a great deal of information about life and its relationships.

As one of the oldest shared elements of most life on earth, DNA acts as a partial record of our ancestries and our ties to one another. It contains patterns that are passed from generation to generation, recording in their alterations and preservations the histories of fusions and fissions that have characterized life from its first humble appearances as much as 4 billion years ago. By comparing different organisms' DNA sequences, we can reconstruct minute details of their lineages on the great map of life. In short, the structure of DNA and the way this structure is passed from generation to generation enable us to use it, alongside the fossil record, to create a map of the relationships and histories of all living things.

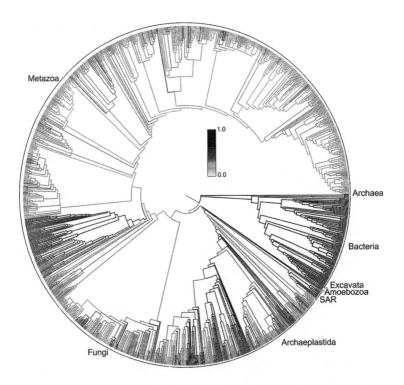

Figure 1. The grand panoply of life, in which humans are but a teeny twig. Source: C. E. Hinchliff, S. A. Smith, J. F. Allman, et al. 2015. "Synthesis of Phylogeny and Taxonomy into a Comprehensive Tree of Life." *Proceedings of the National Academy of Sciences of the United States of America* 112, no. 41 (2015): 12764–69.

Looking at humans as part of the cluster of backboned animals, we see that we share something like 98 percent of our DNA with chimpanzees, slightly less with monkeys, around 80 percent with dogs, and about 60 percent with chickens. It turns out that the first part of the answer to who we are is that, biologically, humans are squarely in the midst of the order of mammals we call "primates." Even a superficial look at all the primates (including us) reveals the family resemblance. What then does being a primate have to do with the capacity for belief?

Humans as Primates

If you stand in the middle of a large group of macaque monkeys early in the morning, the first thing you notice is the activity: feeding, grooming, playing, fighting, resting against one another, exploring, watching each other. To your left you might see a "matrifocal cluster": an older female, her sister, her daughter, her daughter's daughter, and their one- and two-year-old offspring playing around them, leaning into them, grabbing bits of food from their hands. The three generations of adult females groom one another, running their hands though each other's fur, calmly sharing the physical space and creating a social bridge across generations. A few meters away, an old male watches the cluster groom and keeps his eye on two young adult males who have newly begun to challenge him in their quest for social position. He sighs, rises, and saunters over to the females, lying down in front of them to signal an invitation. The oldest turns away from her daughter and begins to groom the old male. This activity, the primate way of being in the world, is social through and through. The deep roots of primate sociality are a key source of the human capacity to believe.

As animal behaviors go, primate sociality is particularly complex[2] and different from other forms of gregariousness. Several bird species flock together. Wildebeest congregate in herds that travel hundreds of miles. Fish of all sorts school, swimming and feeding as if they were a single composite organism. But primate groups, distinctively, feature collections of individuals with a range of personalities, competing interests and shared histories of cooperation, conflict, trust, and manipulation. To be social as a primate requires a particular intensity of dynamic relationships. The infrastructure of primate sociality sets the stage for the cognitive and social resources necessary for belief capacities to emerge.

Some species of mammals and a few birds have independently evolved complex social lives similar to the primates. We see these patterns in whales and dolphins, in wolves and dogs, and also in the viverids (especially the meerkats) and their cousins the hyenas.[3] But in primates, especially in monkeys and apes, sociality is more than the sum of its parts. The combination of a very extended and intense mother-infant bond, long life spans, substantial cognitive capacities, and the emergence of strong and diverse personalities with an anatomy that includes agile grasping hands, an upper body that frees the arms and hands when sitting down, color vision, and a lack of highly specialized morphology for combat (such as claws or spikes) facilitates the emergence of primate sociality. The primates' assemblage of physical traits and cognitive and social capacities gives rise to certain possibilities for seeing the world, for complex behavior, for an intense inquisitiveness, and an ability to manipulate objects and other group members in fascinating ways. The world would undoubtedly be very different if orcas (killer whales)[4] had thumbs and legs for terrestrial movement. But they don't. We do.

The capacities inherent in primate bodies and minds enable a diverse range of relationships between individuals. Bonds among genetic kin are expected, but friendships between those not genetically related also play a central role in many primate lives. Alliances and coalitions—short- and long-lived—social manipulations, deceit, and changing relations between individuals as their social landscapes shift are all part of primate sociality. For primates, the social landscape is the stage for existence, and in evolutionary terms, each primate is shaped by—and also shapes—its social ecology.[5]

In making these assertions, I'm drawing on more than sixty years of intensive research on multiple primate species. For apes such as the chimpanzees, gorillas, orangutans, and gibbons, and for monkeys—especially baboons, macaques, capuchin monkeys, muriqui monkeys, and vervet monkeys—we have multiple generations of data from many sites around the world.[6] Over decades of work, thousands of researchers have given us thousands upon thousands of hours of observation of primates both in captivity and in the wild, as well as intensive investigations into their behavior, physiology, genetics, and morphology. And we are primates. We belong to a lineage for whom social interactions, social landscapes, and social connections with our kin, friends, and even enemies are central to our lives and an integral part of our evolutionary story. More than almost any other animals, primates have a wide-ranging capacity to respond to challenges with behavioral innovation.

To be specific, the human capacities for cooperation, collaboration, and deep social reliance on and commitment to others—the critical social and cognitive resources underlying the capacity for belief—are rooted in our histories as primates. This is not a novel idea. Darwin hinted at it, and many contemporary primate researchers have done important work demonstrating that this is indeed the case.[7]

Yet there is more to learn from primates. In addition to the capacity for complex social lives, heightened social cognition, and behavioral flexibility, there is another piece that offers additional insight into the evolutionary infrastructure of our capacity to believe. That additional piece is primates' ability to have an aesthetic sense and the experience of awe—critically necessary for the development of transcendent experience.[8]

A few years ago I collaborated with some colleagues at National Geographic, the group that designs and implements critter-cams[9]—cameras attached to various animals to allow us humans to see what the world looks like from the animal's point of view. We placed cameras on two species of macaque monkeys, one in Singapore and one in Gibraltar. The camera attached around the neck and sat just below the chin on a swivel such that whether the individual was sitting or walking we got more or less an orientation and image that represented the animal's point of view. These glimpses into the daily lives of the primates are amazing, at least for primatologists. We saw feeding, grooming, fighting, playing, and the monkeys' entire social world acted out in front of us from their perspective. But there was something else too. Only a few instances, not enough to develop any real scientific analyses ... but something.

In Singapore, a male wearing the camera climbed down from a tree and approached the large six-lane highway dividing him from a forest patch and an enormous fruiting fig tree. He leaped onto the pedestrian overpass and climbed onto its roof, where he began his jaunt across the bridge, presumably headed to the fig tree. But halfway across the bridge, the monkey stopped. Now remember, we are reviewing this footage on a monitor after recovering the camera and are seeing not the monkey himself but his point of view. He spends a few seconds moving toward the fig tree and then turns toward the edge of the walkway roof, overhanging the six lanes of speeding traffic. He approaches the edge and stops, looking over . . . we see the image of speeding cars rushing past in both directions. Then he sits up and the frame is filled, like a painting, with borders of forest, a center of highways and speeding traffic, and high-rise flats in the background. It's a stunning view. He sits motionless, watching this panorama for several minutes. Then the image shifts. The fig tree flies back into view and gets closer and closer.

In Gibraltar, a female macaque named Sylvia wore the camera for us. She provided a dynamic and exciting day of footage: Sylvia's hands as they groom her young son, her spats with a male and a female, her hands foraging for flowers and grasses along the peninsula's steep cliff faces. But in late afternoon she stops. She sits high up on the western face of the rock of Gibraltar, overlooking the strait from which we can just make out the outline of Jeb el Musa, Gibraltar's twin mountain, on the north coast of Morocco. We, the three researchers, are mesmerized by the beautiful scene, but Sylvia is not done. She shifts her body and head slightly, and we gasp. The new scene she has framed is even more breathtaking. The light shimmers off the meeting of the Mediterranean and the Atlantic, North Africa rises magnificently on the horizon, and the edges of the rock frame the left side of the image. Sylvia sits for many minutes, taking in the scene before moving on.

These are anecdotes. But they are hard to forget, and there are others. Many of my colleagues who have spent months and years watching primates report moments when a primate seems to be enraptured by a scene, possibly enjoying a purely aesthetic experience. The primatologist Geza Teleki wrote of a particularly stunning experience of meeting up with a few adult male chimpanzees and sharing with them the silent viewing of a sunset.[10] Others have similar stories, and many are equally powerful.

We know enough about primate visual systems and neurological processing to know that we and they are seeing more or less the same thing. Could there be some aspect of visual experience that resonates across species? Might the emotion of awe be older than humanity? I wish had an answer, but I don't. All the same, this idea suggests an interesting possibility about the primate perceptual and experiential baseline that humans expanded upon, especially when we consider the anthropologist Mel Konner's suggestion that the sense of wonder is the hallmark of our species and the "central feature of the human spirit."[11]

What we know about the primates suggests that the core structures of primate sociality and social flexibility and their distinctive social and cognitive resources offer insight into the evolutionary basis of the human capacity for belief. I also suggest, without irrefutable supporting data but with no lack of interesting possibilities, that being primates also implicates us, via our particular visual, cognitive, and social systems and histories, in the capacity for the aesthetic and a sense of awe.

All of this is not yet the explanation I have promised, but it offers a critical component of the story: our shared history as primates is the baseline. But humans are the strangest primate. Our entangling of the social and the imaginary, the economic and the political, the compassionate and the cruel, our capacity to believe, is not simply an extension of primate trends. The human story is not just the primate story.

Evolution is about continuities and discontinuities. As lineages diverge from common ancestors they retain many similarities, but the differences, the particular evolutionary paths, are what make a lineage distinctive. Our split from our closest primate cousins occurred some 8 to 10 million years ago when we began our independent, highly distinctive evolutionary journey. The specific lineage to which we belong is called the hominins, and a second aspect of the answer to who we are, and why that matters for belief, lies in the bones of these ancestors.

HOMININ EVOLUTION IN A NUTSHELL

Rarely does a distant ancestor look exactly like its descendants. The earliest hominins were more like the common ancestor they shared with the other African apes, the chimpanzees and gorillas, than like we are. When they branched off from their closest relatives (the ancestors of the African apes) sometime between 8 and 10 million years ago, our ancestors were medium-bodied, mostly quadrupedal, with strong grasping hands and feet, and a fairly large brain relative to other primates. Their upper bodies had the shoulder blades pushed to the back, allowing full rotation of the arms, and a flattened chest.

The early hominins represent a critical transition, from habitually quadrupedal movement, mostly in trees, to habitually bipedal movement on the ground. The shift from a four-legged to a two-legged primary mode of movement radically changed how our ancestors interacted with each other and the world. Standing up, in the trees and on the ground, changed the use, and eventually the capacities, of hands and bodies, and opened a new panorama of communication with gestures and motions. This morphology enabled our ancestors to carry more items from one spot to another and to experiment with new manipulations of stone and wood, as well as each other, combining hands, eyes, faces, and minds in novel and trailblazing ways.[12]

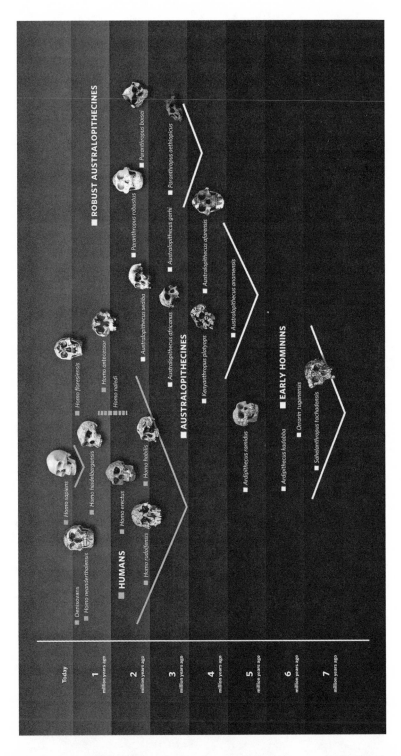

Figure 2. The timeline of hominin and human evolution. Source: NHM Images © The Trustees of the Natural History Museum, London.

We have a number of fossils of these earliest hominins. Despite what each of their discoverers has claimed, there is no scientific consensus on which, if any, are the ancestral members of our specific lineage (*Homo*). Ranging from the piecemeal evidence of *Sahelanthropus tachadensis* to *Orrorin tugenensis* to the more ample finds of members of the genus *Ardipithecus* and the earliest Australopithecines (*Australopithecus anamensis*), these early hominins display a few traits that are relevant to our story. They are all found in Africa. They were all small-bodied compared to modern humans, and their brains had the size one would expect for apes of their size. They were bipedal, but not in the same way we are, and they spent a lot of time in trees. They ate largely plant-based diets. Importantly, all were food for an array of large predators. And they persisted.

Between 3 and 4 million years ago, multiple hominin lineages spread across eastern and southern Africa, and the most likely candidates for our specific lineage were among them. We have hundreds of fossils of the hominin we call *Australopithecus afarensis*, whose populations, many argue, formed the basis for all subsequent lineages. *A. afarensis*, small and best known from the famous female representative called "Lucy," had a brain sized much like its immediate ancestors and had long arms—good for moving in the trees—but legs and a pelvic girdle, indicating frequent bipedal movement on the ground.[13] They were the prey of large cats, eagles, hyenas, and other predators.

Nevertheless, they did well enough for their populations to diversify across eastern and southern Africa. By around 2.5 to 3 million years ago we see multiple new lineages of hominins, including our own. Evidence suggests this was a great time of evolutionary experimentation, with many variations on the hominin theme emerging.

The evidence we have about the social world of *Australopithe-cus afarensis* is generally taken as the baseline for what our specific lineage initially had to work with.[14] They moved as a group and ranged over large areas. Most likely they did a lot of mutual grooming and a good deal of socializing that may or may not have involved various kinds of gestural and vocal communication (but not language). Given that they had many predators both on the savannas and in the trees, predation was likely a significant selection pressure, and these hominins must have had some form of antipredator strategy. It may have been simply to run away, or it may have involved individual or group defense. Either way, avoiding getting eaten was likely a very important element in the life of *A. afarensis*.

Most (but not all) researchers believe that *A. afarensis* exhibited significant sexual dimorphism (males being much larger than females), yet both sexes' canine teeth are of similar size, an atypical pattern for most primates. Generally, when a primate species has large males and small females, the males have much larger canine teeth, adult males tend to be socially dominant over females, and males and females live overlapping but largely distinct lives. Similar-sized canines are most often found in primates where males and females are approximately the same size and live in small groups, with male and female roles overlapping extensively and males doing a lot of caring for the young. Humans today show a strange mixture: males about 15 percent larger than females, same-size canines in both sexes, and a lot of diversity in our social patterns. It is very likely that this deviation from the primate standard in form and social style—or some version of it—characterized our Australopithecine ancestors as well.[15] Mixing up expected patterns of behavior seems to be a deep part of our history.

One of the main fossil finds of *A. afarensis* (called "AL-333") consists of seventeen individuals at a single site and may offer at least an anecdotal illustration of these hominins' lives. Just over 3 million years ago in the Afar region of Ethiopia, nine adults, three teenagers, and five young children of the species *A. afarensis* traveled across the mostly open grassland dotted with clusters of trees. Wherever they were headed, they never made it.[16] The way the fossils are spaced indicates that all seventeen died at about the same time, but not from a flash flood or other natural disaster. Some have suggested that the group consumed poison, but it is more likely they were attacked, probably by very large cats or other predators.

We don't know how large the actual group was—there may have been more than seventeen—but it probably was not much larger. We also know that predators, even when three, four, or five of them are hunting together, make one or a few kills in a group of prey, then stop to either eat them there or else take the carcasses away to eat them elsewhere. This means that most or maybe all of the group of *A. afarensis* may have stayed to assist the few first attacked, and in the end all died. If this is what happened, it represents a kind of extreme group cooperation in the face of intense danger that is not at all common among animals, or even most primates. Such a tragic event might be early evidence of hominin groups bonding in more cohesive ways than are seen in most other animals—even at the expense of their own lives—a distinctive social and cognitive investment in one another.

Let me offer two more quick notes about the Australopithecines that are central to understanding our lineage.

In 2010 at a site in Ethiopia called Dikika, researchers discovered marks on animal bones that are about 3.4 million years old, the oldest evidence of animal butchery with stone tools ever found.[17] But they found no actual tools at the site. From the animal fossils, it is clear that flesh was cut and scraped off the bones and that stones were used to either break the bones or loosen the meat. The most likely candidates for having done this are *Australopithecus afarensis* or one of the other two hominin species around at the time, *Kenyanthropus platyops* and *Australopithecus deyiremeda.*

In 2015, researchers working at a site called Lomekwi 3, near Lake Turkana in Kenya, found the earliest known stone tools; they are 3.3 million years old.[18] The tools are mostly cores of stone with flakes that were deliberately chipped away by hitting the large stones against an anvil stone to produce specific shapes and edges. We do not know who made them, but given the age and location it's most likely that *Australopithecus afarensis* earn the prestigious title of the first stone toolmakers in the history of our planet.

Between 2.5 and 3 million years ago, the first evidence of members of our own genus, *Homo,* show up in the fossil record.[19] At this point we know they inherited not only our primate capacities for complex and interconnected social lives, and possibly for an aesthetic sense and the experience of awe, but also the hominin patterns of bipedality and a greater use of their hands. In the hominins, the relations between morphology, behavior, and intense group cohesion were greater than in most other primates. In addition, shortly before we see the first sign of the distinctive human lineage, evidence suggests that our deep ancestors developed the ability to see stones as potential tools and to actively reshape rocks into new forms. To this day, stone toolmaking is a feat no other lineage on the planet has ever mastered.

THE BEGINNING

Today, after an 8- to 10-million-year run, we have emerged from great diversity as the only hominin left standing. We are the strangest primate and the strangest hominin. Just looking around ourselves (wherever on the planet we might be) and taking in all that surrounds us—thinking about the amazing buildings and histories, our diverse lives, wondrous discoveries, theoretical contributions, and marvelous prose and scholarly insight that has been and is being produced, it is absolutely clear we are more than simply a clever and complex bipedal primate.

CHAPTER 2

What Makes Us Human?

APPROXIMATELY 2 MILLION years ago our direct ancestors endeavored to succeed in a landscape filled with large predators and competitors for food and shelter. What did a cluster of medium-sized, hairless, fangless, hornless, clawless hominins have? They had each other and the glimmering of a new way to shape, and be shaped by, the world. In order to understand why we believe, we must delve into the distinctive story of human evolution.

WHAT IS A HUMAN?

Humans are mammals, primates, and hominins, but we are also something distinct. We can philosophize about our place in the world and share it with one another. We can experience existential crises, and we regularly assume that certain aspects of life include a transcendent component. Even an evolutionary explanation of what makes us human, therefore, must share some connection to these distinctive facts of humanity. Such considerations are mainstays of many philosophical and theological explanations for what makes us human, but those explanations largely draw on other sources of knowledge than evolutionary science. An evolutionary description of human distinctiveness has to include a measurable or otherwise discretely identifiable set of parameters that reliably classify different clusters of beings.

So what is it that makes us human in the evolutionary sense?

We can start with the idea of *species*. Species definitions separate organisms by shape, size, genetic sequences, and other variables that arise during their evolutionary histories. There are today about twenty-six ways to classify species.[1] These constellations of anatomical and genetic measures can identify you and me as human and not dolphin or elephant or chimpanzee. But what makes us human, even evolutionarily, is more than a constellation of morphological traits or genetic sequences. Organisms of any species are not simply museum specimens to be measured and catalogued.

Measuring an eagle's wings, head, and talons, or assessing its genetic sequences, allows us to differentiate it from a sparrow or an ostrich, but this approach does not tell us what it is to be an eagle for the eagle itself. Living organisms are not static assemblages of traits. They do things in the world that are not reducible to morphology or genetics. If what organisms *do* is critical to understanding what they are, an evolutionary definition must encompass not just an organism's materiality (morphology, genetics, etc.) but also its behavior, its way of relating to the world.

We could say that humans are human because we have complex social lives, our children take a long time to develop, and different groups of us have different social traditions in how we act, eat, mate, and use tools. This is also true, however, of orangutans, chimpanzees, gorillas, orcas, dolphins, elephants, and many other socially complex mammals. To understand, from an evolutionary perspective, what makes a particular kind of organism distinctive, we need to know not just its form and various details of its behavior but also the way that it exists in, and with, the world. To paraphrase anthropologist Tim Ingold, we must understand how it "becomes."[2] We need to be able to describe its *umwelt*, the way it perceives and experiences the world.

Philosopher, semiotician, and biologist Jacob von Uexküll illustrates an umwelt by describing the life-world of a tick.

From the egg there issues forth a small animal, not yet fully developed, for it lacks a pair of legs and sex organs. In this state it is already capable of attacking cold-blooded animals, such as lizards, whom it way-lays as it sits on the tip of a blade of grass. After shedding its skin several times, it acquires the missing organs, mates, and starts its hunt for warm-blooded animals.

After mating, the female climbs to the tip of a twig on some bush. There she clings at such a height that she can drop upon small mammals that may run under her, or be brushed off by larger animals.

The eyeless tick is directed to this watchtower by a general photosensitivity of her skin. The approaching prey is revealed to the blind and deaf highway woman by her sense of smell. The odor of butyric acid, that emanates from the skin glands of all mammals, acts on the tick as a signal to leave her watchtower and hurl herself downwards. If, in so doing, she lands on something warm—a fine sense of temperature betrays this to her— she has reached her prey, the warm-blooded creature. It only remains for her to find a hairless spot. There

she burrows deep into the skin of her prey, and slowly
pumps herself full of warm blood.

The tick's abundant blood repast is also her last meal.
Now there is nothing left for her to do but drop to earth,
lay her eggs and die.[3]

The blind, deaf, tiny tick lives, perceives, acts, and reacts in
the world in distinctive ways. Modern ecological and evolutionary parlance uses the term "niche" to describe, scientifically, the
umwelt of an organism.

WHAT IS A NICHE?

Over sixty years ago, biologist G. Evelyn Hutchinson famously
described the niche as the dynamic multidimensional space
in which an organism exists—the totality of the biological and
material factors that make up an ecological space occupied by
a species, how it lives in the world. Naturalist Joseph Grinnell
defined the niche as an organism's ecological role, how it lives in
its habitat. Today, evolutionary biologist David Wake and his colleagues identify the niche as a species' structural, temporal, and
social context. It includes space, structure, climate, nutrients,
and other physical and social factors as they are experienced and
restructured by organisms via the presence of competitors, col-
laborators, and other agents in a shared environment.[4]

Niches are dynamic, not static, and they are the nexus of any
evolutionary story. For an evolutionary answer to the question of
what makes us human, more than just measures of our bodies and
behaviors, we need to understand the structure and dynamics of
the human niche.

THE HUMAN NICHE

The human niche is the spatial, ecological, and social sphere that includes all social partners, perceptual contexts, and ecologies of human individuals, groups, and communities and the many other species that live with and alongside humans. Humans can occupy multiple subgroupings across space and time and can share cognitive, social, and ecological bonds even without being physically close. It is within their niche that humans interact with, modify, and are modified by their social and ecological environment throughout their lifetimes.[5]

The human niche is where we share social and ecological histories; where we create and participate in shared knowledge, and social and structural security; and where our development across the life span occurs. But for humans, the niches we occupy also include the perceptual contexts—the ways in which our structural and social relationships are perceived and expressed through behavioral, symbolic, and material aspects of the human experience.[6] Anthropologist Terry Deacon describes this aspect of the human niche as the "great ubiquitous semiotic ['meaning-making'] ecosystem in which we develop."[7] In this niche, which is simultaneously ecological, material, imagined, perceived, and constructed, meaning matters and is evolutionarily relevant. So belief is certain to play a key role.

Anthropologist Polly Wiessner and I recently offered an overview of the human umwelt, in the manner of von Uexküll's description:

A fetus is formed via the interactions between the genes and developmental processes, laying the baseline for body and behavior. In the womb, it is exposed to environmental factors such as diet and stress that shape its development and can set off epigenetic change. After birth an infant may be strapped to a cradleboard, cuddled by the father, or nursed by a number of caretakers with an impact on the physiology of both caretaker and infant. From early on, children have the dispositions to charm investment from community members and acquire the roles and rules of society. They begin to embody the skills to negotiate challenging physical and social terrain. Even basic perceptions such as smell and color are mutually shaped physiology and cultural experience. Growth and maturity are often ushered in by complex rites of passage, with social selection pressures shaping reproductive chances and outcomes and what those processes mean to the individual and the society. Humans develop in community. Adults carry out economic enterprises in roles built over generations of history. They acquire ideological outlooks that guide their motivations, goals, and loyalties. Humans learn the rules of cultural institutions while individual agents push the limits, bringing about game changes that alter social structures and make history.[8]

Given the vastly complex interlacing of biological, social, and cultural forces, understanding the human niche and its evolution is a daunting challenge, especially given that in contemporary evolutionary theory, simplistic and reductionist notions of what organisms are and how they came to be are now seen as insufficient. The biologist Kevin Laland and his colleagues perfectly summarize contemporary evolutionary understandings:

"Organisms are constructed in development, not simply 'programmed' to develop by genes. Living things do not evolve to fit into pre-existing environments, but coconstruct and coevolve with their environments, in the process changing the structure of ecosystems."[9]

To understand how niches are initiated and constructed, and how this process relates to belief, we need to be clear on how the processes of coconstruction and coevolution work.

Contemporary evolutionary theory is much more than "survival of the fittest," the simple process of more "fit" variants leaving more copies in subsequent generations (which is how most people think of natural selection[10]). The current state of knowledge about how the process of evolution works is called the extended evolutionary synthesis (EES).[11]

We can recognize five clusters of evolutionary processes.

Biological variation. The first cluster explains the source of biological variation, the core matter of evolutionary change. Genetic mutation introduces genetic variation (variant forms of DNA sequences), and these new sequences interact with the inner workings of cells and other systems that affect how DNA functions (called "epigenetic" systems). These interactions are drawn into the range of developmental processes (growth and interaction of tissues, organs, and bodily systems). Suites of interactions, from DNA to epigenetics to the development of bodily systems, produce biological variation in organisms that may be passed from generation to generation. This variation is the material basis on which evolutionary processes work.

Shaping the variation. The second cluster acts on this variation. It includes the process of natural selection, as proposed by Charles Darwin and Alfred Russel Wallace and modified by more than a century and a half of research.[12] Natural selection shapes the details and patterns of variation from generation to generation in response to specific constraints and pressures in the environment. Variants that are better fit to a given environment leave, on average, more offspring than those variants that are less adept at dealing with the specific pressures in their environment. Genetic variation is also affected by the movement and mating of individuals (called "gene flow"), which actively mixes and redirects the patterns of genetic material. Finally, in a process called "genetic drift," chance events also affect the distribution of genetic patterns and sequences.

Organism-environment interaction. The third cluster focuses on the dynamic interaction between organisms and their environment, including what we call "niche construction." Niche construction is the building, modifying, and destroying of niches via the mutual interactions between organisms and their immediate environments, which can sometimes change both. The effects of niche construction can shape the patterns and intensity of natural selection and other evolutionary processes (from cluster 2) and create ecological inheritances in the form of altered niches passed down across generations. These outcomes can change the evolutionary pressures on subsequent generations. This third cluster is about how organisms both shape and are shaped by the world around them.

Plasticity of bodies and behavior. The fourth group of processes are the plasticity and flexibility in behavior and bodies that emerge from the interactions of everything in clusters 1, 2, and 3. This plasticity and flexibility (or lack thereof) manifests in the lived experiences of most organisms and shapes the patterns and the structures, and the production of evolutionarily relevant outcomes.

Modes of inheritance. Finally, the fifth cluster reflects the fact that there are multiple modes for the inheritance of evolution-arily relevant variation from generation to generation. Biological and behavioral information and patterns that affect the processes in clusters 1 through 4 can be passed along via DNA sequences (genetic) and non-DNA processes (epigenetic), as well as through behavioral transmission. Humans can also inherit symbols, perceptions, and beliefs, and these can be evolutionarily relevant.

This overview of the way theorists now categorize evolutionary processes should make clear that evolution is much more than "survival of the fittest." It is a complex bundle of interactions by which populations of organisms both shape and are shaped by many aspects of their environment. These patterns produce niches that are coconstructed, inherited, and altered across space and time.

The development of the human niche is the process by which the bodies, minds, and lives of the members of the genus *Homo* have changed. The fossil and archaeological records reveal increasing brain sizes and changing neurobiological capabilities, and myriad changes to skulls, legs, feet, and other material elements of bodies and lives across evolutionary time. Yet a common concern for paleoanthropologists investigating this "human niche" is when to start calling the members of the genus *Homo* "human." If becoming human is connected with the capacity to believe, then drawing the line that defines "human" also implies locating the point where we think belief became significant (or possible).

There is serious contention over how many species are represented in our lineage and how they relate to one another. We have no consistent species definition for past hominins, especially members of genus *Homo*, nor any agreement among researchers on how to develop one. There is, however, near unanimous agreement that by about 1 million years ago, all populations of hominins on the planet belonged to genus *Homo* (all other lineages having gone extinct) and that there has been a lot of morphological variation in populations of the genus *Homo* over the last million years.[13]

For the last half century or so, paleoanthropologists and archaeologists relied on two major developments as having "identified" the emergence of contemporary humans: behavioral modernity and anatomical modernity.[14] Anatomical modernity was the date at which fossils that look like the bones of contemporary humans show up in the archaeological record. This used to be around 100,000 years ago, then about a decade ago it was pushed to 180,000 years ago, and in the past few years it has been pushed again by new fossil finds to around 300,000 years ago.[15] The concept of behavioral modernity relies on a "human revolution" in which significant cognitive changes created a disjuncture between *Homo sapiens* (true humans) and our closest hominin relatives (other members of the genus *Homo*, usually classed as Neanderthals and *Homo heidelbergensis*, or just labeled "archaic" species of *Homo*).[16] Until recently this shift was identified by the appearance of "symbolic" artifacts in the archaeological record. If an ancient population made symbolic items, it was cognitively human. The timing of this event has been variously pegged to the appearance of cave art and carved figures (ca. 65,000–40,000 ya [years ago]), to specific types of stone tool technologies (ca. 75,000 ya), and to the use of ochre for engravings and pigments (ca. 80,000–180,000 ya).

It should be obvious from the changing dates that this search for anatomical or behavioral modernity is misguided.[17] Every time we expand the data set for our past we have to change our criteria for the dividing line between true humans and other humanlike members of genus *Homo*. We now know that the package of morphological traits we once labeled "anatomically modern" shows up in varying forms, in varying places, in different populations of *Homo* across Africa and Eurasia over the last 300,000 to 400,000 years. We also know that almost every type of material evidence, of "symbols" or other indicators, that we considered a sign of uniquely modern human cognition also shows up associated with populations that do not have modern human morphology. Sometimes these signs show up before any populations with modern human anatomy exist.[18]

Becoming human was and is an evolutionary process, not a creation event. There is no clear line in the last half million years where we can state with absolute certainty, "This group of fossils is our one and only direct ancestor." The evidence suggests, on the contrary, that many populations in genus *Homo* contributed to contemporary humanity.[19] As our knowledge of the human fossil and archaeological records expands, and as our ability to develop more robust genetic and evolutionary models increases, we see that the last half million years is a time of major transitions in genus *Homo*'s biology and behavior. Rather than seeing these changes as radical shifts in material cultures that indicate new and distinctive cognitive capacity, and thus a new species, we can see them more as a process. Contemporary *Homo sapiens*' (our) success was not due to a set of specific morphological or material adaptations but rather a suite of changes in how members of the genus *Homo* perceived and interacted with the world and each other: a change in the human niche. The available evidence suggests that the elaboration of behavioral practices, patterns, and processes over the last half-million years indicates a particular expansion in the complexity of the human niche.[20] A core part of this process was the fuller development, and eventual application, of the capacity for belief.

For the vast majority of human history, the total number of individuals in our lineage (genus *Homo*) was fairly low, and most populations probably had no regular contact with more than a few others. We know that populations did interbreed, at least a little, and that most eventually went extinct.[21] Starting between 300,000 and 400,000 years ago we begin to see evidence of more frequent connections among populations within and across regions, a scenario that expands greatly in the last 100,000 years. Between about 300,000 and 25,000 years ago, our own morphotype (bodies that look similar to today's humans) becomes dominant, and by about 25,000 years ago humans who look like us are the only members of our genus on the planet.[22]

Today, all humans are not only the same species but the same subspecies, *Homo sapiens sapiens*. We have more in common with one another genetically than any mammal with a global distribution—even more than many with very limited distributions. In fact, every human alive today has much more in common, genetically, with every other human than do chimpanzees living in eastern Africa with those living in central Africa.[23] Contemporary humans are enormously successful, fantastically widespread, and extremely genetically similar.

What we call our ancestors, or even ourselves, while important, is not what explains why we can believe. The answer is in the processes of how we got to where we are today: the construction of the human niche.

CHAPTER 3

Constructing the Human Niche

I F HOW we create, occupy, and modify the human niche is a large part of the answer to what makes us human, then a substantive part of *why we believe* lies in the patterns and events that characterized the process of human niche construction across the longue durée of our evolutionary history.

Over the last 2 million years, the human lineage underwent specific morphological changes, alongside significant but less easily measurable behavioral and cognitive shifts as it forged a new human niche.[1] From a group of fangless, clawless, naked, bipedal, apelike hominins, our lineage developed into makers of stone tools, controllers of fire, and producers of cave art—and then became the constructors of towns, cities, and empires, and ultimately a principal force in the global ecosystem.

Our success is linked directly to the emergence and development of a distinctive human niche, a novel umwelt. And we have material evidence for many of the major landmarks in this development.

The human niche developed slowly at first, picking up the pace in the last 400,000 to 300,000 years, expanding dramatically in the last 125,000 years, and exploding in complexity over the last

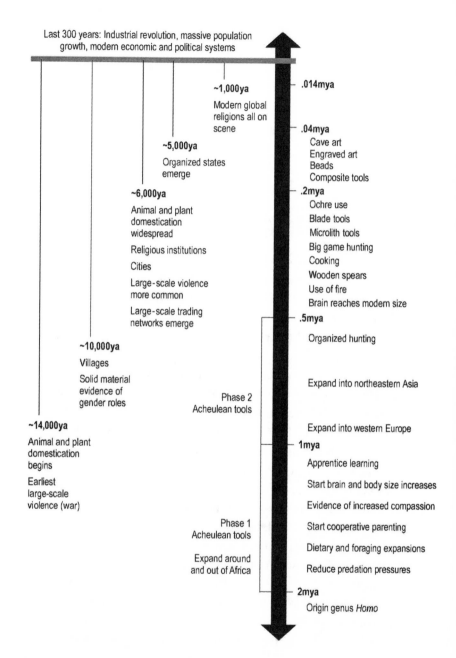

Figure 3. A timeline of the development of key elements in the human niche.
Source: Fuentes, "How Humans and Apes Are Different."

12,000. Many populations of our genus contributed behaviors, genes, lifeways, and other traits to our current incarnation. We will not worry about what to call these populations. Other populations of genus *Homo* were humans, not exactly like us, and many eventually became extinct, but their physical, physiological, and behavioral legacies contributed to the underlying processes of contemporary humanity—to our niche.[2] This dynamic mode of becoming, shaping, and being, shaped by an increasingly human niche, forms the core of our capacities for being human, and for believing.

For the purposes of this book, we can divide the development of the human niche into two basic parts: the initiation of the niche, and the construction (but not completion) of it.

INITIATING THE HUMAN NICHE

As Kevin Laland and his colleagues remind us, "Organisms are constructed in development, not simply 'programmed' to develop by genes. Living things . . . coconstruct and coevolve with their environments, in the process changing the structure of ecosystems."[3] A core challenge we face in examining human evolution is to identify the critical events and processes that facilitated the coconstruction and coevolution of the traits, bodies, and ecologies that enabled the development of the contemporary human niche.

This is immensely complicated. We know that the feedback loops, interfaces, and dynamics involved in human evolution, past and present, are extremely complex. For example, archaeologists Matt Grove and Fiona Coward reviewed the core range of available data and came up with a "simplified" diagram highlighting more than fifty specific traits, behaviors, and patterns that mutually affected one another in the processes of constructing the human niche during the last 2 million years.[4] For the purpose of examining the emergence of our capacity for belief, however, I focus first only on a few key patterns and traits on which a vast majority of researchers agree played pivotal roles in the initiation of the human niche. Its roots lie in the connections between nutrition, bodies and brains, caring for young, avoiding predators, and making stone tools, in the time between roughly 2.3 million and 1 million years ago.

The earliest members of our genus walked on two legs and had longer arms and fingers than we do, remnants of their recent ancestors' continued use of trees. But their hands were comparable to ours, with substantial dexterity. From quite early on, our ancestors diverged, even if but slightly, from the other hominins. Their brains were a bit more dynamic, and their capacity for making stone tools a bit more complex.[5]

Making stone tools involves not only the capacity to envision alternative and useful shapes in a stone but the ability to develop and share the processes for creating those shapes by modifying the stone. And not just any stones. There is good evidence that by 1.8 million years ago, early humans were choosing certain types of rocks (as much as ten to twenty-five pounds of them!) and transporting them up to five to ten kilometers before shaping them into tools.[6] This toolmaking capacity, including the source selection, material transport, and sharing of manufacturing techniques, was most likely possible because a suite of social learning processes was more developed in our lineage than in those of other hominins.[7]

We have substantial evidence that our ancestors were preyed upon by numerous large carnivores.[8] But over time we see fewer fossils of our lineage that have clear evidence of predation. Predation was a problem our ancestors appear to have solved, at least partially, by making themselves a less ideal target. They likely did this by expanding their capacities for better social communication and coordination, and for predicting predator movements. These actions would have made them less available to predators compared to other sorts of prey. Our ancestors deployed their imaginations and distinctive cognitive and social resources not just in shaping stones into tools but in reshaping the predation pressures affecting them.[9]

This suite of social and communicative behaviors led to increased capacities for coordination, eventually enabling groups of *Homo* to scavenge predator kills more effectively, increasing the amount of animal protein in their diet. Now able to extract more from the landscape, they expanded their foraging patterns and enhanced the nutritional returns even further.[10] By just under 2 million years ago, early members of the genus *Homo* were starting to mix in near-shore aquatic resources such as turtles and catfish, and possibly roots and tubers into their diets, in both cases using stone tools to help them extract these resources.[11] There are many benefits in expanding one's dietary range: it allows one to get better nutritional value with less effort, to avoid competition with other species, and to obtain food with less risk. Perhaps most importantly, it gives one a variety of fallback options when food is scarce. All of these advantages open up new possibilities for innovations in social and cognitive processes.

At the heart of some early *Homo* groups' success was a mixed reliance not just on their bodily capacities and social cohesion, which they shared with most primates, but on their emerging capacity for behavioral innovation, developing new complexes of behavior that interacted with their ecologies and developing minds in new ways. They began to combine social and cognitive experiences and create new ways to think about and act on the world around them. Enhanced stone toolmaking came about through experimentation, imagination, collaboration, and many failures leading to improved learning. Their knack at predator avoidance helped develop communication abilities and heightened their skills at prediction and snap decision making. Expanding their dietary resources enabled novel ways to share those practices intensively and extensively, adding to their nutrition and their overall tool kit. All of these enabled our lineage to develop a level of communication-based social learning, a cooperative and collaborative intensity of information transfer that surpassed those of other hominins. This was the start of the human niche.[12] Our early ancestors blended distinctive cognitive and social resources, histories, and experiences in the context of expanding experiences and imaginations.

Two critical changes necessary to developing the human capacity for belief occurred at this time. First, as social lives became more complex, it took longer for children to learn how to be effective adult members of their groups. Second, as diets became better and more diverse, our ancestors became better able to feed the expansion of their and our most costly organ: the brain. These two patterns, in combination, initiated the evolution of a longer maturation process, which facilitated longer brain growth and more neurological and social development after birth. *Homo*'s brains became anatomically more responsive to their environment—they grew increasingly dynamic neural pathways in response to experience—creating greater learning capacity and enhanced abilities to imagine and to translate those imaginings into social and material reality.[13]

This critically important extended childhood period begins early in our lineage—but at the cost (already high in other hominins and apes) of making their highly dependent offspring *even more* dependent for a longer period of time. This places extraordinary pressures on the mother. To address this pressure, our ancestors capitalized on a component of their emerging niche: their increased capacities for communication, cooperation, and social coordination. Redirecting these capacities toward their new challenges, early *Homo* became cooperative caretakers of their young. More than any other species, they developed a pattern of extended brain development, expanded caretaking, and the immersion of infants and children into a world of intensive cognitive and social stimulation.

How this pattern emerged was brilliantly outlined by anthropologist Sarah Hrdy[14] and later modeled and described by many other researchers.[15] The behavioral changes came first: our ancestors developed a system of cooperative care for the young in which males, siblings, and other group members all likely offered some degree of assistance, be it carrying infants, helping care for toddlers, and feeding (when the infants could take solid food). The system eventually led to physiological changes that primed all humans for broad-scale caretaking—we see the evidence of this evolutionary history today in male and female bodies and physiologies, and in the behaviors and material remains of contemporary and past societies.[16] This system of cooperative caretaking lessened the burden on the mother and expanded the social landscape and learning possibilities for young and old.

Ours is a thoroughly social and cooperative niche, linked to augmented capacities to collaborate, share information, and rely on the manipulation of materials outside our bodies to create new solutions to the challenges of the world. These changes both enabled and drove the growth in complexity of genus *Homo*'s lives, adding cognitive resources and opening new possibilities for creative exploration and imagination. Expanding innovation and communication and changing minds and bodies were precisely the elements needed for developing the capacity for belief.

THE CONSTRUCTION (BUT NOT COMPLETION) OF THE HUMAN NICHE

Between 1.8 million and 100,000 years ago our lineage spread across Africa, much of Eurasia and both East and Southeast Asia. By 70,000 years ago, *Homo* was in Australia, by 15,000 to 20,000 years ago in the Americas, and by 2,000 years ago throughout the islands of the Pacific.[17] Across this period we see evidence for ecological, behavioral, and technological expansion of the human niche. Core aspects of this expansion included more complex interrelations between groups, enhanced foraging, the mastery of fire, the increased presence of meaning-making (called "semiosis"[18]), and increasingly complex tools. All of these factors had a role in shaping *Homo* cognition and social organization.

ACHEULEAN TOOLS

A novel type of stone tool technology, called the Acheulean, emerges and becomes progressively more complex between 1.5 million years and 400,000 years ago. The story of this technology offers insight into the further development of the infrastructure for the human capacities for belief. Acheulean technology requires substantially more training to produce than earlier tool types. Its tools are more complex in their structure and refined in their use, and they enable an even wider array of food processing and material manipulation. Most interestingly, by about 500,000 years ago we find that many of the classic form of Acheulean tools, called "hand axes," are being crafted to an extreme level of symmetry, more than is necessary for function—as if this were an aesthetic choice. These tools are beautiful, with some made from multicolored rock and others with fossil shells left embedded in the stone. In some places, Acheulean tools appear piled in clusters across the landscape, as if they were made and cached, or possibly being used as geographically or culturally significant markers. Acheulean tools are both functional, used to tackle the challenges of the world, and imaginatively created, reflecting something about, and for, their makers.[19]

Recent work by a number of scholars demonstrates that this creativity in tool production had a broad impact on the capacity for flexibility in behavior, diet, and habitat manipulation. It also affects brain function. Anthropologist Dietrich Stout and colleagues[20] have shown through innovative experiments that the efforts of teaching and learning involved in creating these tools can facilitate novel connections across neurobiological systems that are likely related to a restructuring of the cognitive system and to developing what the neurobiologist Michael Arbib[21] calls "the language capable brain." Such a brain plays a central role in our ability to develop complex imaginative and linguistic frames. It develops in concert with the *Homo* social and material landscapes, giving *Homo* stronger capacities for mental imagery, intentionality, and innovation.

By 200,000 to 400,000 years ago, tool kits have diversified still more and start to take on distinctive regional and structural variation. There is evidence of specialized tools for large game hunting, as well as a wider range of materials (including more wood and bone), and the pace of change continues to increase. We also start to see clear evidence of composite tools, with two or more parts bound together.[22] The ability to imagine an assemblage of different types of objects combined into a single tool opened up even more possibilities for increased neurobiological innovation and connectivity and other morphological and behavioral changes.

By this time, our lineage is locked into a lifeway that involves the extensive collection and deployment of external materials that we shape, meld to our desires, and use to alter the world. *Homo* extends cognition and action into the world by expanding the capacities and confines of our own bodies. In the human umwelt, our niche, our lineage becomes at least partly "cyborg"[23] hundreds of millennia before iPhones and artificial limbs. Also around this time, we start to see stones and other material transported across longer distances than in the past (frequently tens of and occasionally more than 100 kilometers), indicating that at least in some places, trade or exchange between relatively distant populations is emerging.[24] Genetic evidence also suggests that such exchanges and connections were increasing. Today's patterns of deep interconnectedness and exchange of ideas between distant human groups begin here.

The behavioral and physiological infrastructure that evolved alongside cooperative parenting and caretaking enabled the development of heightened compassion—another key factor in our capacity for belief. Archaeologist Penny Spikins has documented a wide range of fossil evidence for care of elderly, injured, and infirm individuals.[25] This is not to say that *Homo* did not also fight and kill each other. We have examples of what appear to be lethal injuries caused by other *Homo*.[26] But instances of compassion and caretaking are more frequent and more striking. We have fossil evidence of aged individuals with no teeth who must have been helped by other group members, maybe to the point of chewing their food. Individuals with serious injuries, disabilities, and severe developmental diseases all appear to have been helped by others to heal and live far longer than if they had been left to themselves. No other species, to our knowledge, exhibits such an extensive pattern of care for its infirm and disabled. This capacity is thought to be one of the key reasons why our genus was able to spread so widely, across so many new environments, and collaborate so effectively to develop novel ways of reshaping the world. Compassion became a central feature of the human niche, and remains today as a core feature of many human belief systems.

Cruelty can be a central feature as well . . . but that story comes to the fore in later chapters.

One of the most critical factors in the construction of the human niche was the recognition and eventual mastery of the power of fire. There is contention about when the first populations of genus *Homo* began to use fire regularly. Some put it before a million years ago, others more recently, and the data are equivocal. Regardless, nearly all researchers agree that sustained use and control of fire is found across a majority of *Homo* populations by 400,000 to 300,000 years ago.

The use of fire made meats and plants much more digestible, gave new access to the nutrients of roots and tubers, and led to experimentation with new ways to prepare food. Fire was used to combat, displace, and control other animals, chasing bears and saber-toothed cats out of caves and keeping hyenas and leopards away from the camps where *Homo* slept together at night. With fire *Homo* found they could heat the items they already used, altering their structures and capacities. Heat enabled better and more efficient shaping of rock, hardened the points at the end of wooden spears, and showed *Homo* that seemingly solid material, like ochres or other minerals, could be made soft, even liquidlike.

Fire opened new doors for experimentation, expanding how *Homo* could manipulate the world around them.

But most importantly, fire allowed *Homo* to throw off the constraint of the day-night cycle of light and activity. No longer did nightfall force them to huddle in the dark, silent and wary. Gathering around the fires at night they could continue the day's work, or spend time communicating, possibly sharing ideas and thoughts. Even before full-blown language, fire may have facilitated the emergence of a very human tradition of describing their days, their experiences, and maybe even their dreams. Fire as a tool and as a source of light and heat reshaped our niche and opened up new pathways and opportunities for social and cognitive interactions.[27] It inspired *Homo*'s imagination and manipulation of the world, opening up the space we eventually filled with belief.

Not surprisingly, in this same period we also see increasing evidence for the creation and use of objects whose meaning is not necessarily tied to their material qualities—what many like to call "symbol" or "art." These are the items that most researchers, and much of the public, identify as hard evidence of belief.

Biological anthropologist Marc Kissel and I recently compiled and published the largest database of all known potentially "symbolic," or meaning-laden, artifacts in the early archaeological record.[28] They range from vaguely human-shaped modified stones and evidence of lines and simple geometric designs etched into shells and bones, to the manipulation of minerals for use as pigments and glues, and finally to evidence of personal adornment, such as beads and the use of paints on bodies and tools. Over the last 200,000 to 400,000 years of our history, the human niche developed and expanded the realm of imagination and creativity across populations and groups of genus *Homo*. Making meaning and altering the world, not just to serve specific functional ends but to represent shared ideas and imaginaries, became a central feature of daily lives. As anthropologist Terry Deacon[29] suggests, this is when our lineage solidified a ubiquitous semiotic ecosystem as a central facet of our niche. Nothing in the human niche is without meaning, and most of that meaning is created or modified by humans themselves.

At this point, many components of our capacity to believe—augmented cognition and neurobiology, intentionality, imagination, innovation, compassion and intensive reliance on others, and ubiquitous meaning-making—were present, if not intrinsic, in the human niche. This niche opened new horizons and structures, neurobiologically, ecologically, and behaviorally, that laid the groundwork for the contemporary human capacity for, and practice of, belief.

THE HUMAN NICHE AND BELIEF

We have just traveled at a breakneck pace though nearly 2 million years of our own history. Evolutionary histories are always complicated, ours especially, but no matter how many times I immerse myself in our story I am constantly amazed.

For clarity and continuity, let's come back to the point of this book, draw together the core processes that emerged as part of the human niche, and relate them to *why we believe*.

First, the patterns and processes of social cognition, the human perceptual landscape, emerged from the processes of toolmaking, foraging, caretaking, the control of fire, the creation of symbolic materials, and the ecological expansion of humans across the planet. This ongoing dynamic, the feedback between neural and behavioral plasticity, laid the neurobiological, cognitive, social, and ecological foundations in human populations for a capacity for belief.

Second, the ratcheting up of social and ecological complexity, combined with increased interactions among populations over the last 200,000 to 400,000 years, created opportunities for the connections and exchanges that enabled shared beliefs, and eventually belief systems, to emerge.

Finally, the last few hundred thousand years offer material evidence for an increase in, and eventual ubiquity of, meaning-making in human populations. The members of the human niche became infused with a capacity for imagination and innovation. This capacity entailed the emergence of two significant patterns: first, imagining novel items, representations, and materials and either making them or altering other things to become them. This capacity appears in a limited form in other animals but becomes permanently and ubiquitously part of the human niche. The second pattern draws on the first: over the last few hundred thousand years of our history, as part of our intensive communicative abilities, humans formed the capacity for creating explanations of widely observable phenomena such as death, the behavior of other animals, weather, or the sun and moon. They did not, for example, simply connect clouds, thunder, rain, and floods, they also developed explanations for why these things happen. This capacity is what anthropologist Maurice Bloch[30] refers to when he asserts that over evolutionary time, humans went from being socially complex transactional beings (like most social mammals) to groups of organisms who exist simultaneously in both transactional and transcendent realities, and who use imagination and belief to reshape themselves and the world around them.

Biological anthropologist Ashley Montagu encapsulates this

core facet of the human niche (and here I paraphrase his famous quote):

> The environment humans make for themselves is created through their symbol using ability, their capacity for abstraction. The symbols, the ideas, are created in the mind ... but the human animal learns not only to create them, but to project them onto the external world, and there transform them into reality.[31]

This is our capacity for belief, and it is the product of the evolution of the human niche.

This capacity is not simply an "emergent property" or something ephemeral floating above the material reality of being human. The ability to believe is part of the human system in the same way that fingers are part of our arms and hands. Fingers are core aspects of human anatomy, modified over evolutionary time to dramatically expand our options for interacting with the world and each other. The limbs to which fingers and hands are attached have been shaped and altered over evolutionary time so that their ends contain structures and capacities that enable them to do much more than they once could. Our capacity for belief is similar: a core aspect of the human that is critical in the human ability to engage with and shape the world.

By 25,000 years ago the human niche was well constructed. Yet some critical structural, ecological, and perceptual aspects affecting how and why we believe *as contemporary humans* had not yet fully emerged. These include domestication, communal architecture, larger communities, changing notions of identity, politics, and commercial systems. Along with these factors comes the emergence of property and inequality, and of course large-scale violence and warfare. These elements enabled more diverse relations, more types of conflict, and greater coordination and cooperation—and set the stage for contemporary human beliefs and belief systems.

CHAPTER 4

Animals, Plants, Buildings, and Pottery

DOGS, CHICKENS, cows, pigs, apples, mandarins, peaches, potatoes, squash, grapefruits, kiwi and chilies; smallpox, tuberculosis, plague, coronary artery disease, high blood pressure, obesity; towns, cities, temples, markets, courts of law, governments, armies, wealth, poverty—none of these things were part of the human experience before the most recent phase of our history. Many did not even exist for 99 percent of human evolution, only coming into being, or into relevance, in the last 15,000 years.[1]

These new elements affected the patterns and infrastructure of human societies and ecologies in critical ways. Basic patterns of land use, nutrition, social grouping, and our relationships with a diverse array of plants and animals were altered, creating new ways for humans to perceive and evaluate one another. Humans also began to change how they created memories of place, histories, and ideologies.[2] These changes are fundamental to the social and ecological structure of everything that came afterward.

We know that the human niche, our way of being in the world, took shape over the last 2 million years. The feedback between behavior, physiology, neurobiology, bodies, and ecologies starts slowly, reshaping what our ancestors looked like and how they acted, but then picks up speed as we draw closer to the current day.

At first, heightened abilities for cooperation and collaboration made humans better at avoiding predators, making and sharing stone tools, widening their dietary options, communally raising their young, and reshaping local ecologies. Well before 1 million years ago, populations of genus *Homo* had spread across Africa and much of Eurasia, exploring new landscapes and developing novel approaches to deal with the challenges of diverse environments. In meeting these challenges, they grew increasingly able to draw on a range of cognitive and social resources, as well as histories and experiences. They used their emerging imaginations to think beyond the immediate present and to develop mental representations of the world and their lives. The capacity for belief began to play a role in the human niche.

By 400,000 years ago populations of *Homo* were using fire to alter stones, wood, food, and the day-night cycle. They were hunting large game and developing intricate tool kits. By 200,000 years ago our ancestors were combining materials from plants, animals, and minerals in increasingly complex new forms, and by 100,000 years ago they were painting themselves, making jewelry, and pushing into the more northern landscapes. By 70,000 years ago they could cross open water and were creating images and using color on cave walls to tell stories about the world. Basic beliefs and maybe even some rudimentary belief systems became central to their everyday existence and to the long-term success (and failure) of human populations.

By 25,000 years ago our ancestors came to look, and in many ways act, as we do today. But the human niche—especially the most powerful aspects of human belief—had not yet come into its contemporary manifestation.

In the last 15,000 years or so, the magnitude, diversity, and impact of humans' relationships with other species, the places we inhabited, and ourselves, transformed bodies, societies, and the global ecosystem at a pace outstripping everything in the previous 2 million years. It is in this most recent period of human evolution that we see the emergence of significantly more complex, and decidedly multispecies, human communities, economies, polities, and belief systems. Domestication, storage, and new modes of social structure enabled the rise of new ideas and practices of property and identity that, in combination with expanding patterns of inequality, created radically different landscapes of human experience. These landscapes set the stage for what we now term the *Anthropocene*,[3] the age of humans, and the niche in which we now exist—a niche that includes an elaborate array of novel options for compassion and cruelty, belief, hope, and despair.[4]

THE NEW WAY OF BEING HUMAN

Domestication is generally seen as the modification of a plant or animal species in a way that accentuates the traits beneficial to human use. In the case of wheat or rice, for example, it was the reshaping of the grains so that they became larger, more nutritious, and less likely to drop off their stalks, making the plants reliant on humans both to collect the grains and to propagate the next generation. In the case of goats or cows it was manipulations to develop smaller and tamer breeds that stayed around human settlements. These new members of human settlements learned to communicate effectively with humans (at least understand core instructions given by them) and grew fast, providing meat, milk, bones, skin, and horns for human use.

But domestication is not simply control over the bodies and lives of other organisms. It is an ongoing process of niche construction in which two or more species reshape one another. Much popular and academic literature assumes that domestication came about when some human populations, pushed by environmental stresses, gave up their roaming ways as hunters and gatherers to focus on a few specific species to cultivate, remodel, raise, and consume. This focus on necessity and stress probably does not capture most instances of domestication.[5] Recent overviews by archaeologists Melinda Zeder and Bruce Smith[6] reveal a different, more interesting story that informs us in critical ways about the human niche.

By about 15,000 years ago, some human populations began to develop larger settlements with more cohesive residences of 100 to 200 inhabitants, either joined together or clustered over a short distance. These villages were often located in resource-rich ecosystems. While they may or may not have been permanent settlements, they were occupied for much of the year, and the human populations exploited the diversity of plants and animals around them for food, buildings, tools, art, and clothing. Wherever humans roamed, at least some populations developed relatively settled communities with resource management systems for the surrounding landscape that were passed from generation to generation and were shared among multiple groups living in the same region.

In addition to the selective use and harvesting of wild plants and animals, these resource management systems included a habitual reliance on behavior that tied human lifeways to the lifeways of those plants and animals. This was not yet domestication as we know it today, but these relationships between humans and other species began a mutual process of interreliance that left marks in the soils, in the species compositions of forests, in the patterns of bones of animals, and in the bodies and materials left behind by the human populations.[7] The processes of domestication arose out of recursive interactions between humans and others, in which the other species became more than simply food or tools for the humans. They became central components of our physical and perceptual realities, morphing into integral parts of the human niche, much as stone tools and fire had done earlier in our evolutionary history.[8]

A hunter/forager's relationship to animals differs from that of a domesticator. For the forager, there can be a deep mutual interaction with other species, with humans shaping much of their activity to those of the animals and the other species responding to the pressures emanating from human predation. For example, across millennia human groups in northern parts of the world structured their lives around the migrations of herds of caribou or reindeer or buffalo, who in turn shaped their ranging and feeding behaviors based on human predation. But with domestication, relationships move beyond predator-prey interactions to a much more direct and intensive mutual reshaping of bodies and lifeways. The investment in one another and the processes of relationships become more integrated and interreliant.

Over generations, domestication created strong feedback loops that reshaped humans and other species in distinctive directions.[9] Many of the changes in domesticated species that increased human returns on investment in care and guardianship also made these organisms more reliant on humans. Think of our substantive and complex entanglements with dogs, cows, sheep, chickens, rice, wheat, and potatoes. With domestication, deep relationships across species lines have become a central component of the human niche.

ANIMAL DOMESTICATION

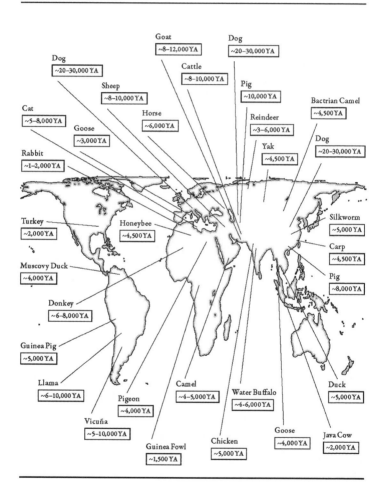

The first material evidence of patterns of relationships that we can clearly call domestication is dated to more than 12,000 years ago and becomes more common as we approach modern times.

PLANT DOMESTICATION

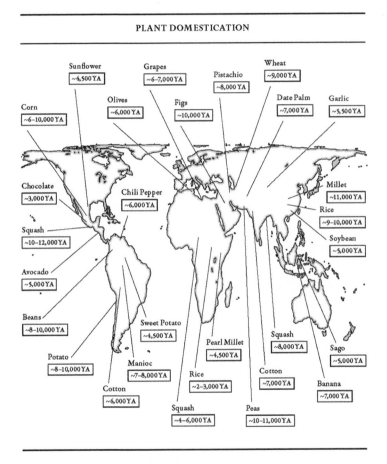

Figures 4a. and 4b: Maps and timing of plant and animal domestication. Source: Agustín Fuentes. 2017. *The Creative Spark: How Imagination Made Humans Exceptional.*

The evidence reflects significant changes in the shapes of seeds, fruits, roots, and plant bodies; in the legs, bone density, and sizes of animals; and in the bodies and lifeways of humans.[10] These relationships are found across the planet. Domestication has no single point of origin. Rather, it is an outcome of the human niche over many populations, ecologies, and geographies. Also, the specific animals and plants that entered into these relationships have moved as humans moved, traded, and commingled around the planet. Today you can sit in a café in New York and eat a meal of potato soup, a cappuccino, a cheese and chicken salad, and a coconut mango flan for dessert whose ingredients originated between 2,000 and 10,000 years ago in the South American, African, Southwest Asian, and Mediterranean regions. You are thus benefiting from the innovations of hundreds of ancestral populations of humans and their mutual engagements with thousands of animals and plants.

To give a hint of how this happened and how it influenced why and how we believe, let me provide two brief examples: the domestication of dogs, and that of rice.

Dogs

More than 12,000 years ago, at the site of Uyun al-Hammam in what is now Jordan, a group of humans dug a grave and laid two bodies into it, one on top of another. One was that of an adult woman, the other an adult fox. They added a few tools and some red pigment, and then covered the bodies. At nearly the same time, at the Natufian site of Ain Mallaha in what is now Israel, a group of humans stood among the oak and pistachio trees and laid a young woman in a shallow grave, placing a deceased dog next to her head. They laid her hand on the dog before covering them both and sealing the grave. By at least 12,000 years ago, dogs were clearly important in the beliefs of some human populations.[11]

Humans and dogs were the first domestication event, and the relationship between them began not as an intention but as an overlapping and fusing of niches. About 25,000 years ago two species, gray wolves and humans, were successfully spreading across the northern hemisphere. Both were highly social mammals living in complex communities. They had in common a strong loyalty to their groups, communal care of the young, and an effective hunting skill set that relied on cooperation. Human communities and wolf packs often came into conflict, especially over kills. But sometime around 25,000 years ago, some wolves started following hunting groups, scavenging the more frequent human kills and becoming constant figures on the periphery of human communities.

At first, the wolves avoided direct confrontation, remaining just out of reach. But the longer the wolf packs remained around humans, the more likely it was that a few individuals began to venture closer to the human encampments for easy food (garbage and waste) and the protection of the nightly human fire. The humans would have likely driven the wolves away. Over time, if the wolves remained persistent, just out of reach but always nearby, the humans may have become accustomed to their presence. The wolves would have begun to follow the human groups as they moved about and remained around the camps if the humans settled in for a season or two. These humans would have noted that there were benefits to having the wolves around. The wolves made noises if other predators approached the camps at night, and they shadowed the humans when they hunted, sometimes driving out smaller prey that the humans could capture and eat as they looked for larger game. Finally, it appears that on multiple occasions when abandoned or injured pups or young wolves ended up in human camps, the humans nurtured them rather than killing them.[12]

It is at this point that relationships started to change.[13] Over time, initially in eastern and central Eurasia, humans and wolves started to see one another as neighbors and hunting partners, and eventually as friends. The wolves that grew up around human communities started to transform. Our ancestors noticed differences in the wolves' personalities that affected their ability to get along with the human community, how adept they were at communicating with humans and following human cues, and even how they interacted with human children. Some showed more attachment and more responsiveness to human signals and less inclination to fight humans over kills—even bonding tightly with particular individuals, following them around, and keeping watch over them. These wolves transferred their pack allegiance to the humans, and the humans developed ways to shape it. Before long, humans recognized that behavioral training with pups produced the best companions. As our ancestors began to selectively spend time with the most human-friendly pups, the behavior and bodies of the wolves changed: they became dogs.

The traces of this story show up in the bodies and DNA of dogs and in the archaeological remains of human activities. By 15,000 to 20,000 years ago, remains of human encampments contain bones that are wolflike but show signs of domestication, smaller and less robust, even a bit more puppylike. We also know from contemporary research on domesticating another canid, foxes, how such processes might have been undertaken: specific caretaking behaviors prompted by humans' perceptions as to what constituted a better "wolf" slowly changed the bodies and behavior of wolves into those of dogs.[14] Human relationships with the emerging dogs also shaped the way humans structured their lives, and possibly how they saw the world.

Today the human-dog relationship spans the globe, with dogs playing central roles in many communities and societies, and contributing significantly to billions of humans' lives. Dogs form key parts of the cognitive and social resources of human populations, and they help shape our experiences and imaginations. For many humans, our relationships with dogs (and other domestic companions, such as cats) influence the way we experience attachment, friendships, and emotions: dogs are central components of many human belief systems.

RICE

We should never think that plants lack agency. Their lives can be dynamic, and their histories can play key roles in the shaping of human bodies and lifeways.[15]

Today *Oryza sativa* is the staple food for nearly 50 percent of earth's population. We know it as "nasi," "arroz," "mǐfàn," and "rice." Domesticated several times over the last 9,000 years across Asia, and about 4,000 years ago in Africa, this small grain plays a fundamental role in the lives of humans and other animals.[16] The model for the ancestor of contemporary rice is another species in the genus *Oryza*, *Oryza rufipogon*, a sinewy grass with hard red seeds that thrives in swampy areas of Asia—edible but tough, weedy, and hard to control.

Around 8,000 to 12,000 years ago, human communities in what is today China were becoming more settled and were improving ways to store food. These communities selectively harvested the grains of early wild rice and possibly removed some of the competing plants around it. The challenge for humans trying to gather this rice is that the stalks produce few grains, and as these ripen, they fall off the stalk and land in the swampy water or on muddy soils, where they are eaten by birds and other animals. But there is variation: some grains hold on more tightly than others.

Sometime before 9,000 years ago, human populations living around the Pearl and Yangtze Rivers began to selectively collect stalks, removing those whose grains fell off easily and letting the tougher ones remain. Humans began increasing the representation of the tougher stalks by weeding out the weaker ones. With plants whose grains held on, humans could remove the top of the stalk, carry it to their camp or village, and remove the grains by hitting the stalks against a hard surface. They gained more of the plant's nutritional benefit this way, with only minimal loss of grains to animals or germination.

There is a genetic mutation in rice that toughens the connecting tissue between the grain and the stalk and makes it harder for the grains to drop off. It shows up naturally now and again but disappears quickly in most populations of wild rice, as plants that carry it have trouble dropping their grains and reproducing. But by about 5,000 to 7,000 years ago, the mutation shows up in most strains of rice associated with human communities. Humans reshaped the genetics of rice by favoring specific plants.[17]

Since each stalk of rice has so few grains, feeding humans requires many stalks. If they are not able to reproduce by themselves because humans are keeping all their seeds, then humans must not only gather the rice but also replant it so that new stalks will emerge. Even a community of just a few hundred people needs thousands and thousands of rice stalks.

A reliance on rice demands a large community, substantial coordination and planning, and some division of the labor involved in collection, processing, and replanting. Human domestication of rice reshaped the plant's body, genes, and life cycle, but at the same time, reliance on the plant changed human labor practices and social structure. Human group composition, individual roles, storage needs, and annual cycles all became attuned to the needs of the rice as well as those of humans.[18] Human daily experience and perception were substantially altered by the investment in rice. And the same outcomes emerged from tending other plants. Farming activities bent toward future outcomes could be uncertain and would invite theorizing about plant behavior, weather, and interactions between plants and other organisms, including explaining the inevitable variability in outcomes. Acting toward future productivity, fully investing in plants as central to a society, demands a certain belief in the effectiveness of these acts and a sense of understanding of the way the world works.

This pattern is true for most relationships with domesticated plants and animals. The needs of domesticated organisms and our reliance on them have substantial implications. Corn, beans, wheat, potatoes, cows, goats, horses, pigs, llamas, and many others require social and ecological contexts, structures, and commitments from the humans who rely on them. The plants and animals we have incorporated into our niche affect how we live, how our bodies grow and function, and how we see ourselves and the world around us.

Domestication is a reciprocal and dynamic pattern of mutual influence[19] that reshaped critical factors in the human niche and now greatly affects how we experience the world and what we believe about it. As humans committed more deeply to certain animals and plants there developed a greater and greater intimacy, and the likelihood of producing explanations of animal and plant behavior grounded in ideas about agency, awareness, and so on increased. Such explanations could easily go beyond simple observations into ideas of anthropomorphism, sentience, and even attribution of other kinds of awareness. Domestication reshaped how humans believe.

Humans Make a Commitment

Domestication emerged in many populations around the planet, but it did not do so in a vacuum. Around the same time that domestication was spreading, many human populations made a particular set of commitments. These included the use of storage for accumulation and preservation, intensive investment in specific locations (sedentism), and the emergence of large, permanent building projects.

Storage and the Emergence of Material Inequality

If we think about where we put food, dishes, clothes, and things that have special meaning to us, it becomes very clear that storage is central in our lives today. It turns out that the development of storage was one of the most important factors driving the emergence of the contemporary human niche. Small storage containers of hide, leaves, or empty shells are most likely quite ancient in the human lineage. The carrying of surplus food and other items extends hundreds of thousands of years into our past. But we have little archaeological evidence to document this.

We do find evidence, over most of our genus's history, for the caching of items in reasonably secure locations, but little sign of preservation or accumulation. The differences are important. Accumulation storage indicates an ability to generate surplus amounts of the item being stored, and preservation storage suggests sufficiently advanced technology or insight to develop ways of keeping items beyond their normal period of usefulness—for example, by drying meats, fruits, or seeds and storing them in sealable containers.

Over a million years ago, populations of *Homo* cached clusters of stone tools in specific places to which they would return again and again. But there is no evidence, until much later, that they collected materials or built items specifically for storage. By 80,000 years ago, populations in Southern Africa are using ostrich eggs as water containers and abalone shells to transport and store pigments and glues. By 25,000 years ago, such patterns appear in every human population.[20]

Evidence for the creation of novel items specifically for storage, such as baskets or pottery, starts to appear in the archaeological record about 27,000 years ago, but it is likely that baskets were developed even earlier. Regardless, by 8,000 to 12,000 years ago (around the same time as the initial explosion in domestication), ceramic containers are common at many human sites, and it is with the common use of ceramics that storage and its implications shift dramatically.[21]

Once items of value can be stored, preserved, and accumulated, new types of social, political, and economic relationships can emerge. These new relationships create novel cognitive and social resources, as well as new histories and experiences, and enhance mental representations and imaginations.

The creation of surplus permits a distinct context of perception: the possibility for substantive notions of ownership and the ideology of "property."[22] For much of human history, most groups' social structure contained little role differentiation, and goods acquired by the group were generally shared. Individuals may have had a few personal items for which they would have claimed some form of exclusivity, but not many. Of the few forager groups and small-scale societies that persisted into recent centuries, nearly all show very little status differentiation.[23] In archaeological sites more than 20,000 to 30,000 years old, there is little evidence of clearly differentiated status or roles. A diverse body of data supports the idea that the majority of human populations were relatively egalitarian until the last 12,000 years or so.[24] That is, most resources were shared, with little individual accumulation of, or political control over, goods.

But once storage becomes a prominent part of a population's tool kit, one can see how accumulation and preservation might increase the value of goods and expand the range of options for access and control. Storage creates opportunities for new types of commitments, new perceptions, and new beliefs about ownership and property.[25]

Sedentism and a New Devotion to Place

Another core commitment that shows up in the last 12,000 years is being tied to particular locations, often called "sedentism." Once a population becomes heavily invested in a specific species or set of species as its primary food source, its social and ecological structures are molded by the needs and requirements of those species.[26] This often means committing to a specific place.

In the case of agriculture, this commitment can entail particular residential patterns and a need for a larger labor force. This in turn leads to a division of labor for planting, tending, harvesting, processing, and eventually storing, guarding, disbursing, and replanting. Animal domestication can create new structures of need that human groups must meet, which means addressing their requirements for food, care, and reproduction; protecting them from predators; housing them; and doing whatever processing is needed to extract nutrients and other goods from their bodies. Once formed, these domestication relationships can be disrupted only at the risk of nutritional and social collapse.

These deep investments in space, place, and other organisms lock humanity into a world where intensive obligations to other species are the key to our survival. We are never free of them. It should not surprise anyone that such obligations bring the other species into central roles in the worldview and perceptions of human groups.

Many human populations through the last four or five millennia avoided investing in animal or plant domestication. Yet these forager populations did not live in a vacuum. They often interacted, intermarried, traded, and exchanged with sedentary or domesticatory groups, making such relationships critical parts of their own ecological and social systems. Over the past 6,000 to 8,000 years, the realities of expansion and solidification of domestication, agriculture, and sedentism across human populations had ripple effects that reshaped our entire species, even among humans who did not make those commitments.

Constructing Human Place

The third commitment is material and labor investment in new types of relationships within and between populations. Sedentism greatly expanded a particular form of human collaboration: the building and maintaining of material structures (homes, villages, and eventually cities). Initially, most human-built structures are living spaces for multiple group members. But larger and more diverse structures with multiple purposes eventually emerge. Early villages show evidence of multiple rounds of rebuilding, expanding, and collapsing. They reveal a new kind of relationship between human groups and their locations: permanent residence in human-constructed places. These places provide new landscapes, new structures for human living, and the potential for novel forms of identity, all of which shape human memory by creating more tangible, lasting, and structured material histories.[27] For most of our past, human groups[28] hunted, gathered foods and materials for tools, in and around the same locations for centuries. They would have developed associations with landmarks: specific streams and rivers, hills and mountains in their areas, and these associations were mapped to markers in the surrounding land shaped by ecological and geological forces. Such markers were habitually visited by human groups, inspiring a sense of connection and belonging. But these familiar landmarks are unlikely to have stimulated a direct sense of ownership or of property, as they exist without substantive proactive shaping or management by the people—the landmarks are "already there," part of the world. There were exceptions. Campsites used repeatedly for countless generations, cave paintings, or large piles of stones gathered into specific locations might take on specific meanings due to their intentional and curated presence. In these cases, especially with the cave paintings due to their relative inaccessibility, there could have developed a sense of belonging, even a sense of demarcated possession (this is "ours") for the groups or individuals who made and used them. But even if a group developed some sense of ownership of a set of cave paintings, it was likely different, less intensive, or more ephemeral than the ownership of large-scale permanent residences and the re-creation of landscapes that come with sedentism and domestication.

The longer human groups remained and invested in specific locations, the more important these places became in structuring the possibilities and processes of belief. Archaeologist Ian Hodder and colleagues recently demonstrated that long-term occupation and building in one place creates attachment to the place and facilitates ownership and investment, a specific pattern of history that affects how and what people believe. This work at a site called Çatalhöyük (in what is today Turkey), which was occupied from 9,500 to 7,700 years ago, shows the evolution of particular sets of social practices, building constructions, and ritual and economic relationships. The research tracks over time the material evidence of the actual buildings (sleeping rooms, cooking areas, spaces of worship and of burial) and their contents from an incipient village to a massive collection of residences, communal structures, and shared spaces for hundreds of people. This site offers a nearly 2,000-year window into the ability of humans to build certain ways of living and expand on them. The residents of Çatalhöyük innovated and created new spaces that reflected a melding of the existing material structures, the imaginations of past and present residents, and beliefs about what their group could achieve. The patterns of building, growing, and reshaping at Çatalhöyük offer material evidence of history making[29] during this critical period of transition from one way of being in the world (mobile foragers) to another (sedentary villagers). One of the most critical factors of such history making in this time period, especially in the context of belief, is the creation and maintenance of monumental architecture.

In this terminal Pleistocene-early Holocene period, when so many transitions in lifeways are occurring, we get the first evidence of big, complex human-made structures that are not residences. One of the earliest known examples is Gobekli Tepe, in what is today Turkey.[30] Gobekli Tepe was a large mound site with multiple buildings and perhaps as many as 200 stone pillars, each averaging nearly six meters in height and weighing a few tons. The site was developed and arranged over a few thousand years, starting about 11,000 years ago. It is very impressive.

Building and maintaining this site would have required hundreds of people working in coordination across generations. Yet there is no evidence of permanent residence at the site, which forces us to ask: what was it for? Many have suggested it was a ritual center, possibly a center for some form of organized religious activity. Without a time machine, there is no way to confirm or refute such an assertion.[31] Regardless of its specific function, Gobekli Tepe is a clear example of groups of humans coming together to create a singular place, one they obviously cared about enough to invest significant energy, time, and effort over tens or hundreds of generations. This required coordination, role differentiation, and some form of *belief* that these actions had a greater meaning than the physical nature of the materials.

Gobekli Tepe is not unique. Between about 10,000 and 4,000 years ago, humans in South Asia, Northern and Eastern Africa, Meso- and South America, East and Southeast Asia, and the Middle East started building places, often monumental, that are neither residences nor storage facilities. They appear to have expanded their meaning-laden manipulations of ecologies and landscapes from cave painting, figurine carving, small objects of art, and the manipulation of animals and plants to the massive transformation of spaces and landscapes into wholly human places of meaning—meaning that may have been sacred.

If we take "sacred" to mean sites, places, or objects that are deeply relevant, primordial, and infused with great power in a particular people's belief systems,[32] then we can confidently argue that "sacred" spaces and objects are present deep in human history. However, by 8,000 years ago, humans around the planet are indisputably constructing, maintaining, and using structures such as monumental edifices, walls around towns, plazas and gathering spaces, sanctuaries and altars, and a myriad of built elements that materially reflect complex, coherent, and coordinated belief systems. Human beliefs initiate dramatic new ways of constructing places and altering spaces and inscribing new forms and values into and onto the landscape. By 5,000 years ago, with the frequent presence of monumental architecture and the emergence of urban contexts, we see initial evidence for contemporary religions, political structures, and economic systems.[33] This co-occurrence should not surprise anyone.

The Emergence and Expansion of Inequality

Once human populations became committed to the reciprocal relationships with ecologies and practices we've just reviewed, new patterns of social structure began to develop[34] and introduce novel frameworks for difference and inequality. These opened doors to new ways of structuring our perceptions of the world. It's worth briefly discussing three examples of this process as they relate to belief.

First, across the last 12,000 years or so we see a move from "domestic storage," where goods are stored in living spaces throughout a settlement and distributed among the group more or less equally, to "public storage," where goods are kept in nonresidential structures or areas, such as storehouses or granaries. Public storage needs management, which means centralized control, usually by specialists who run the storage place itself and the movement of goods to and fro. This group of managers constitutes an elite social cohort with responsibilities that set them apart from the rest of the population. Rules and processes of management and distribution also come into play. These are often codified via rituals or other types of ceremonies that create specific roles for the individuals officiating and coordinating such activities. This process creates new landscapes and new social perceptions and relationships.

The reciprocal relationships among storage, surplus, management, and the creation of elites can be seen in many human populations across the planet over the last 6,000 years, as populations increase in size, complexity, and density.[35] Think, for example, of the Sumerians; the Egyptian Pharaonic periods; the Inca, Aztec, and South Asian Empires; the East Asian dynasties; and the Eastern and Southern African kingdoms.

A second important pattern of differentiation that increases inequality and shapes belief is the development of gendered social spheres. "Gender" is how a society designates the roles, expectations, and beliefs it holds regarding the sexes, including reproduction and sexuality.[36] Gender is central in the makeup of belief systems, but gender and gender roles are neither static nor uniform across space and time. Almost everyone would agree that there were gendered differences in the deep past, but they left little material evidence. Today we use differences in grave goods, biological markers of diet or status, or differential skeletal wear or injury patterns to identify gendered behavior in past populations. For example, we see indications of differences in diets between males and females in later farming populations (ca. 4,000 years ago) as measured by differences in their bone chemistry. But evidence of such difference is extremely rare in older samples. There is some evidence from a few Neanderthal sites of differential wear on the teeth of adult males and females, and a few other bits of evidence of gender roles in sites between 20,000 and 35,000 years ago,[37] including examples of different injuries in males and females. However, these cases are few relative to the overall fossil record, and none of the sites show consistent, clear connections to contemporary patterns of gender differences (those we see in many humans today, such as markers of male prominence in economic and political spheres or female association with daily food preparation). While gender must have deep roots, it is not at all clear that early concepts and practices of gender resembled current gender patterns.[38] While much material evidence of contemporary patterns of gender difference does not show up before 12,000 years ago or so, it then becomes clear and common.

With the commitment to agriculture and sedentism, patterns of nutrition changed, and humans were able to supplement their infants' diets with high-value, easily consumed foods like bovid (cows' and goats') milk and grain-based porridges, enabling mothers to wean them at a much younger age. When human mothers nurse they often experience lactational amenorrhea, meaning that the body shuts down fertility while nursing is going on. This enables mothers to produce more milk to assist the infant's growth. Once mothers stop nursing, the fertility cycle kicks back in.[39] With earlier weaning, women begin to have more frequent reproductive cycles. There is substantive evidence that the time between births drops from about five years in forager populations to as little as two years in sedentary agricultural and urban populations.

This shift, combined with changes in residence patterns favoring smaller household units, changed women's mobility and activity and tethered reproductively active women, at least perceptually if not physically, more firmly to the home. In many agricultural and pastoral societies, meanwhile, the nature of some work involved in planting and tending to large-scale crops and managing large herd animals favored size and strength. These requirements led to males becoming allied more with the public sphere, outside the home. This does not mean women were actually weak in these societies. Recent research has demonstrated that many women in this period worked extensively in fields and with tools and had upper-body strength equivalent to or surpassing that of today's female athletes.[40] What is central is that during this period we see consistent evidence of greater differences in how males and females lived, compared to earlier periods. Such changes established a new baseline for perceptions, expectations, and beliefs about gender.

This summary of the emergence of gender patterns is a generalization; a lot of variation exists in exactly how gender roles and norms play out among humans around the planet today. Yet the pattern of there being identifiable differences in behavior reflected in the bodies of males and females (in addition to those directly related to the biology of reproduction) emerges in many places in the most recent phase of human history. There is good physical evidence that, over the last 12,000 years, a trend of increased gender differentiation has led to more stringent social norms, with young males and females trained along different trajectories in order to fulfill these expectations.

The third pattern of differentiation and inequality comes with the expansion of villages and towns into cities. With cities come the diversification and specialization of craftspeople as the needs of human populations grow more complex. This brought differences in training and status, which translated into differences in access to knowledge and social goods, which in turn affected peoples' perspectives. Diverse economies and population structures meant new types of political organization. In many urban populations, elites became more than managers of storage and surplus and began to oversee and designate the movement of people and goods and the distribution of labor across the society. Emerging diversity in economic patterns brings new forms of inequality.[41]

Over the last 5,000 to 6,000 years, patterns of political, economic, and gendered differences, rooted in greater inequality, become central realities for most human populations, a process that lays much of the groundwork for contemporary belief systems. These changes occurred not only within populations but also between them, opening the space for a novel development in the human niche, and one that is intimately connected to belief: warfare.

A New Kind of Cruelty:
The Emergence of Warfare

While humans have attacked and killed one another since the dawn of our lineage, coordinated, large-scale lethal violence between groups is a recent phenomenon. Current work by many anthropologists and archaeologists[42] clearly shows that there is no substantive evidence for regular warfare in human populations before the last roughly 14,000 years. Contrary to assertions by psychologist Steven Pinker[43] and many others, human evolutionary history is not replete with either warfare or Hobbesian barbarity. Nor are we simply peace-loving hominins. There is no doubt, however, that the material record shifts starting about 12,000 to 14,000 years ago, and by 4,000 to 7,000 years ago, evidence of warfare and mass killings becomes a regular element of the human niche.

They key to the capacity for warfare is not simply greater stakes in resource competition. These must be combined with the capacity to turn such conflicts into ideological struggles over beliefs.[44]

For most of human history, lethal violence consisted of relatively rare homicides, revenge killings, deaths from fights over mates, and domestic disputes. These disputes targeted one or a few individuals for specific reasons. But when societies grew larger and more complex, enhanced inequalities and concentrations of political and economic power provided both the incentive and justification for one group to attack another without identifying specific individuals as targets. During the last 14,000 years, humans made the perceptual shift from violence against individuals to perceiving a whole group or cluster of groups as "the enemy." As we redefined ourselves, we also creatively dehumanized others.

New Ways to Believe

The advent of domestication conveyed us to the precipice of the Anthropocene—an epoch in which humans' structure not only local and global ecologies but the very manner in which we interact with them and perceive them. The domestication of plants and animals, the emergence of towns, cities, and communal large-scale architecture—and of large-scale politics and economies, inequality and warfare—all begin in this interim time between our past and our present.

Belief is the capacity to draw on our range of cognitive and social resources, histories, and experiences; combine them with our imagination; and think beyond the here and now to develop new mental representations. Belief is one of the things that gives humans world-transforming power. To believe in the possibility of a new settlement, a new type of society, or a new road to other places is necessary to enable humans to act in ways that can, in the long run, create new realities. Much of the reason we *can* believe the ways we do today rests in changes in the ways societies were structured, interpreted, remembered, and lived over the last 14,000 years. Differentiation in behavior and perceptions, especially in gendered, social, and economic roles, radically expands and diversifies in this phase of human evolution. New economies of scale and content, new realities of inequality, and novel political structures act as templates for patterns and possibilities of belief.

We are well into the Anthropocene. The contemporary human niche emerges from the processes, histories, and patterns we've discussed in these first four chapters. But to more fully understand how all of these elements come together, enabling us to believe particular things, we need to engage the most distinctive development in our lineage: human culture. Understanding what *human culture* actually is, its underlying neurobiology and cognition, and how it works, enables us to truly and effectively understand the importance of *how* we believe, adding one of the final pieces in the answer to *why* we believe.

PART 2
How Do We Believe?

CHAPTER 5

What Is Culture?

IN A SERIES of recent lectures,[1] the former archbishop of Canterbury, Rowan Williams (now Lord Williams of Oystermouth), showed his audience the image of a 40,000-year-old carved figurine with a lion's head and a human body. This figure, he suggested, exemplifies the deep roots of humankind's ability not just to represent things already in the world but to think things anew and turn these new "thinkings" into material realities. The *very fact* of this lion-man figurine, Lord Williams argued, demonstrated a "capacity to depict one such that it speaks of another"[2] and demonstrated that humans in that era were capable of imagining new concepts rather than simply reproducing existing forms.

Anthropologist Ashley Montagu[3] told us more than fifty years ago that the ability to imagine truly new things and ideas and make them into material reality marks humanity as a distinctive sort of being—one capable of belief.

In the first four chapters I explained that humans are a particular kind of primate who manipulates animals, plants, ecosystems, and one another. We saw that over the past 2 million years, humans developed the capacities for intense cruelty and amazing compassion, as well as a distinctive spark of creativity and imagination.

In the preface I defined "belief" as the capacity to draw on our range of cognitive and social resources and to combine them with our imagination. Believing is thinking beyond the here and now and investing to the extent that such thinking becomes one's reality. Beliefs permeate our neurobiologies, bodies, ecologies, and societies. They mediate the whole of human existence.

Chapters 2 and 3 outlined the evolutionary history of the human niche, setting up the context and processes for the evolution and emergence of the capacity for belief. Chapter 4 described how recent changes in our niche—including domestication, property, and inequality—created the social, political, and economic infrastructure for the patterns and contexts of human belief today.

The first section of this book argued that the capacity for belief evolved in the human lineage and is a core part of the human niche. But we have not yet discussed *how* we believe. How do we, in Lord Williams' terms, "depict one such that it speaks of another," or as Montagu wrote, "imagine truly new things and ideas and make them into resoundingly material reality"?[4]

It should not surprise anyone that at the heart of how we believe is the very human reality of *culture*. Answering the question "How do we believe?" with "culture" is not a superficial statement, nor one so general that it is useless. Most people greatly underestimate just what human culture is and how it functions.

Culture is not just the regional flavors of humanity. It is not a merely the specific boundedness, homogeneity, and coherence of any given human society.[5] Nor is it simply a shorthand term for the stuff that humans do and other organisms don't. While social scientists and others often use the word as a "convenient term for designating the clusters of common concepts, emotions and practices that arise when people interact regularly,"[6] this too misses the mark.

Many anthropologists, in fact, discourage their students from using any version of the term "culture" because simplistic uses obscure the reality that the billions of humans alive today make up a "world in which people dwell as a continuous and unbounded landscape, endlessly varied in its features and contours, yet without seams or breaks."[7] I argue, instead, that the word is useful. A system is at work across and within these "features and contours," these distinctive processes of humanity, that evolves as a central component of the human niche. That system is human culture.

Culture is both a product of human actions and something that shapes those actions. It is the context, the framework, the milieu that embodies and gives meaning to our experiences of the world. As the biologist Kevin Laland recently wrote, it is what makes the human mind possible.[8]

How we believe is explicitly an aspect and outcome of human culture. A fuller understanding of how we believe lies in the elaboration of how human cultural processes function. Of course, this definition of human culture is not something that is easily grasped, studied, or examined. In this chapter and the next, I summarize what human culture is, how it emerged, and why it is central to the processes of belief.

What Is Culture?

For humans, culture is at once the ecosystem we navigate and the embodied and perceived processes that facilitate our actions and thoughts.[9] We are born into a world of social and physical ecologies, patterns, institutions, and ideologies that become inextricably entangled with our biological structures and processes before we even leave the womb.[10]

When we consider that humans create and are created by the processes that we group under the term "culture," we must not suppose that this term is somehow opposite to "biology." Culture is not the "nurture" to a biological "nature." Human culture is a dynamic web of significance at the core of human experience, not just a social and historical overlay on our biology. It is a mistake to think our biology exists outside of our cultural experience or that our cultural selves are not constantly coconstituted with, by, and in our biology. "Nature versus nurture" is a fallacy. There are not two halves to being human.[11]

To unpack these assertions, we must clarify what culture is and whether it's truly distinctive to humans. We then need to identify where culture comes from and how it works. With that under our belts, we can go on to examine how human culture makes belief possible.

In 1871 E. B. Tylor[12] defined culture as "that complex whole which includes knowledge, belief, art, morals, law, customs and any other capabilities and habits acquired . . . as a member of society." More than fifty years later, Franz Boas noted, "Culture embraces all the manifestations of social habits of a community, the reactions of the individual affected by the habits of the group in which [they] live and the products of human activities as determined by these habits."

In 1952 anthropologist Alfred Kroeber and sociologist Clyde Kluckhohn, reviewing much of the Western scholarly treatment of culture, extracted 164 different definitions. They synthesized this range of concepts into what they hoped would be the seminal anthropological definition of culture:

> Culture consists of patterns, explicit and implicit, of
> and for behavior acquired and transmitted by symbols,
> constituting the distinctive achievement of human
> groups, including their embodiments in artifacts; the
> essential core of culture consists of traditional (i.e., his-
> torically derived and selected) ideas and especially their
> attached values; culture systems may, on the one hand,
> be considered as products of action, on the other hand,
> as conditioning elements of further action.[13]

While many variations on this synthesis have been put forward,[14] none has expanded greatly on its core notion: culture is
simultaneously the ecosystem we navigate and the embodied and
perceived processes that facilitate our actions and thoughts. In
this description, culture is the central factor of the human niche,
the ubiquitous meaning-laden environment in which we exist.

But given our current state of knowledge, can we state unequivocally that culture, described in this way, is unique to humans?

Do Animals Have Culture?

Over the past half-century, researchers have paid increasing
attention to social traditions in other animals. Social traditions
are behavioral practices acquired through social exposure and
not clearly rooted in a specific genetic or biological cause. Scholars
across multiple disciplines have suggested that culture exists in at
least a few other species, such as other primates and cetaceans.[15]
Given that other animals do not have the same linguistic, cognitive, and technological processes and patterns as humans, we
may need broader definitions of culture than those that anthropologists originally proposed. And if culture facilitates belief

in humans, and other animals have culture, do they also have belief?

Informed by the growing call to heed social complexity in other species, philosopher of biology Grant Ramsey[16] offers us a new definition of culture: "Culture is information transmitted between individuals or groups, where this information flows through and brings about the reproduction of, and a lasting change in, the behavioral traits." By this definition, the cultural information is transmitted behaviorally via social facilitation and learning from others, which endures long enough to generate customs and traditions. If we use such a definition, then what distinguishes humans from other animals is not the presence of cultural patterns and processes but their scale and complexity.

Animal behaviorist Andrew Whiten[17] created a taxonomy of three elements to consider when looking for culture: (1) patterned distribution of traditions in space and time, (2) social learning as central in acquiring these traditions, and (3) the content of the social traditions (actions, materials used, ideas, and so on). Applying this rubric to examine the behavior of humans and other animals, we can see that humans vary more widely than other animals in the specific content of their social traditions, and our patterns of distribution and our uses of social learning are both much more complex than those of other animals. But we are not unique in having culture.

Using both Ramsey's and Whiten's definitions, we can look to our closest relatives, the chimpanzees, for an example of a nonhuman culture. Chimpanzee communities across Central Africa exhibit distinctive social traditions around feeding and foraging. Different communities have their own ways of selecting rocks in order to crack nuts, folding leaves into cups or using moss as a sponge in order to drink from streams, and stripping small twigs of leaves and breaking them to the right lengths for termite fishing.[18] At the site of Golouago, in the Democratic Republic of Congo, researchers have seen chimpanzees carrying small "fishing" sticks over long distances to their favorite termite mounds, where they had left larger sticks during previous visits. The Golouago chimpanzees have also developed a form of sequential tool use. Grasping the large stick with one foot and two hands, they penetrate the ground at the base of a giant termite mound as if digging in a garden. They poke directly down, making a tunnel into which they then carefully insert the fishing sticks, which they have previously altered to have brushlike ends. Wiggling this smaller tool to draw attacks by the termites, they then pull it out and eat the attached termites.[19] This behavior is transmitted from the old to the young as a tradition of this community. To date, we've seen no other chimpanzee groups that do this is in the same way.

Countless similar examples, from sixty years of research on chimpanzees, tell us three things. First, chimpanzees are skilled in using lightly modified sticks, leaves, and unmodified stones as tools. Second, because this skill set shows up in all the apes and in humans, it may be as old as the last common ancestor between great apes and humans. This kind of tool use is a basal part of ape and human capabilities, a shared aspect of ape and human cultures. Third, using tools in this manner is not something an individual just invents in each generation. It is learned through exposure to others in the group, through a kind of social facilitation and maybe even teaching.

That chimpanzees will strip the leaves of a good termite fishing twig and even break it to a desired length, or leave a large stick at a site for future use, shows their capacity to understand that differences in the shapes and sizes of the sticks translate to better or worse tools. This behavior is not confined to primates: we also see it in some monkeys, crows, and other birds. Young chimpanzees, however, can take years to learn how to effectively fish for termites or crack nuts with rocks. Other species can and do use tools, but their patterns and pace of learning are much slower and less information-intensive than even the most basic human processes for doing so. Also, no other species modifies stones to create a new structure, a new tool, and shares the technique for doing so with the other members of her group. This is exactly what began to happen 2 to 3 million years ago, at the very start of our lineage. This creativity and information sharing prefigured a distinctive ability to innovate and create, not common in other animals, that is a central part of the human cultural system.

As interesting as tool use is, there's even more thought-provoking variation among chimpanzee communities. They have many idiosyncratic social behaviors not associated with feeding and foraging that vary across different communities. For example, they engage in a kind of semicoordinated violence that some researchers liken to human warfare.[20] In many but not all chimpanzee communities, males (and occasionally some females) get together and walk the boundaries of their territories in single file. These subgroups are called "border patrols," and the participants are often relatively silent compared to other chimpanzee subgroupings. When these patrols encounter individuals from neighboring communities, they can act violently, sometimes with lethal consequences. Generally, the patrolling group does not attack unless it outnumbers whatever group it encounters. This pattern reflects a certain amount of coordination. A male starts out toward the boundary, maybe quietly hooting, and others slowly fall into line behind him. When they encounter another group, one or two individuals, if they feel the numbers are in their favor, launch into the attack; then the rest of the patrol joins in. Chimpanzee communities also vary in how they greet one another, in their style and pattern of hunting small animals, and in countless details of their daily lives. There is clearly evidence of shared intentionality, a sense of communal goals or objectives, or at least a very strong sense of belonging to their community. There are chimpanzee cultures: dynamic systems of becoming that are central to the chimpanzee niche.

Other species also show evidence of social and cognitive complexity. Cetaceans, particularly orcas, demonstrate social complexity in a range of behavioral patterns that have significant structuring effects on their societies and bodies. For example, styles of hunting are a social tradition inherited via social guidance by the matriarchs, the older females who lead and structure the social lives of the orca pods. These females determine which hunting pattern pods use, and the pods sustain these patterns over centuries, perhaps millennia. These social traditions affect the body size and shape, and even the genetic structure, of different orca populations. Good evidence shows that behavioral inheritance and social traditions of orcas have changed their morphology and genetics.[21] There are orca cultures.

In both ape and cetacean cultures, there is innovation, social structuring of their worlds, and complex cognitive processes implicated in their social traditions. They clearly fit the broad criteria for having culture. A chimpanzee selects a twig, strips it of leaves, carries it—and her young offspring—for thirty minutes, and then uses the stripped twig to capture termites. Afterward she and her young sit, relaxing, fiddling with the twig, and possibly musing over the experience, the sensations, the processes. An older female orca spends decades showing her pod where and how to hunt. Season after season she repeats the behaviors, selecting a particular type of prey, a particular hunting pattern, ignoring others. These behaviors are social and intentional; many would say they are cultural. They require capacity to cognitively innovate, even contemplate.

It is almost certain that chimpanzees and orcas, and no doubt other species, have some form of imagination, and that they draw on a range of cognitive and social resources, histories, and experiences to perceive and navigate their worlds. One might even say that they "believe" in a manner fitting their own umwelts. But they do not share the biological and behavioral patterns of hominins and *Homo* evolutionary history. They do not reside in the niche that we humans do. Human belief, in essence, enables us to see what is not there and to act emphatically as though it was, so emphatically that what is absent is experienced. Human belief can result in the imposing of what does not exist on what does, in some cases so efficaciously that what did not exist comes to exist in the process.[22] To date, we have no evidence that other species share this way of believing.

There is a problem with the inclusive approach that places "culture" on a continuum from less to more complex animal societies: human impacts on the planet are radically different than those of other organisms. Humans do things in ways that chimpanzees and orcas cannot, because we *believe* in particularly human ways.

While by no means arguing that other animals lack complex societies and some form of culture, I want to place human societies and culture in context. To define "culture" as information that is socially transmitted between individuals or groups so that it brings about changes in behavior or tradition makes a good point of cross-species comparison, but it misses critical processes and structures in what humans actually do. These are the things that make up the particulars of the human niche. If we are truly interested in generating insight into how humans are in the world, then direct comparison to other species might not be the right direction to take.

Consider, for example, this cartoon from the artist Dan Piraro.[23]

Figure 5. Cartoon from Dan Piraro illustrating the complexities of human culture. Source: Bizarro © 2009 Dan Piraro. Distributed by King Features Syndicate Inc.

The joke, of course, is that a chimpanzee has turned the tables on Jane Goodall and is studying her and her community just as she studied chimpanzees. The cartoon reflects the classic etho-logical approach to recording the behavior of other animals: direct observation of their social lives. In this manner we have discovered in apes, monkeys, whales, dolphins, elephants, wolves, and even some bird species patterns that fit the definitions of culture as a continuum of the structure, complexity, and intensity of social traditions.[24]

But a little reflection on this method of data collection should make it obvious that we gain only partial insight into human lives by simply observing them. Just watching humans does not give us much insight into why they do what they do, or even how they do it. Why and how humans do what they do cannot be understood simply as observable and shared learned behavior or social traditions. In the Piraro cartoon, why are some wearing shorts and others not? Why are some talking and others not, why do some have pictures on their shirts, why are they wearing clothes at all? And the more interesting and important questions, what are their political leanings, what jobs do they have, how many finished high school, what are their plans for retirement, and so on? The answers are only accessible and understandable in the context of their culture. This is where the distinctions between the specifics of human culture and other animals' cultures become particularly relevant.

The patterns and processes that characterize human behavior and society include many components that are measurably different in scale, impact, structure, and causality from those in most other species. Humans' lived experience includes massive extrasomatic material creation, manipulation, and use—of tools, weapons, clothes, buildings, towns, and much else—and extensive interfaces between us, other animals, and landscapes on scales and with a level of structural complexity unmatched by other organisms: economies, political entities, written histories, and so forth.

The human niche, our ubiquitous semiotic ecosystem[25]—our umwelt—is completely intertwined with language, socially mediated and reconstructed history, institutions, and beliefs. Much of this is not understandable or inferable from simply watching our overt behavior. What humans *do* often has little directly observable, simply interpretable, or even conscious connection to *why* we do it. This is because much of what humans do is structured by what they believe.

To talk about culture in humans is to speak of a different (if somewhat overlapping) suite of experiences, processes, and patterns compared to what we term "culture" in other organisms. Chimpanzees, for instance, have a range of fascinating social traditions and capacities,[26] but they don't have cash economies and political institutions. They don't arrest and deport each other, create massive systems of material and social inequity, change planet-wide ecosystems, build cities and airplanes, drive thousands of other species toward extinction, or write books about all of the above.

But humans do.

If culture is a process that appears in many species across the animal kingdom, then *human culture* is an exponential development of the theme, a niche that makes it utterly different from the rest of the living world.

Tim Ingold, drawing on philosophers José Ortega y Gasset, Deleuze and Gauttari, Ramon Llull and many others, offers ways of thinking about humans and our ubiquitous cultural ecosystem. We should not think of "culture" as a noun, a description. Ingold tells us that "the grammatical form of the human is not that of the subject, whether nominal or pronominal, but that of the verb."[27] We become human constantly and actively with, through, and alongside the human niche, and this process of becoming *is* human culture. When we are thinking about human culture and its relation to belief, it is not sufficient to define *culture* as something humans do or use. We must understand that, literally, it is us and we it.

For humans, culture is the central process in our niche. Anthropologist Pierre Bourdieu[28] offers the term "habitus" for the system of embodied dispositions and inclinations that structures the ways we perceive the social world around us and react to it. This system, he tells us, consists of "structured structures predisposed to function as structuring structures." It is worth rereading that line a few times. It is a rather dense line of argumentation, almost like M. C. Escher's drawing[29] of two hands each drawing the other, but it is accurate. As Kroeber and Kluckhohn note, culture "on the one hand, [can] be considered as products of action, on the other hand, as conditioning elements of further action."[30] For humans, culture is simultaneously the ecosystem we navigate and a ubiquitous, embodied, and perceived experience that facilitates our actions and thoughts at the same time that we use, shape, and create it.

Before outlining specifically how human cultural processes enable belief, we need to understand the evolutionary background of such processes and clarify just when this system we call "human culture" emerged.

Where Human Culture Comes From

Culture cannot fossilize. We can, however, look to our past for material evidence of the human niche in the world. We can use those material fragments to try to piece together when different core aspects of today's distinctively human behaviors and capacities came to be.

From the nineteenth century through today, it has been common among archaeologists and others to peg that start of "true" humanity (and true human culture) to the appearance of the characteristic human behaviors that seem to necessitate language. Historically, researchers traced the origins of human culture to the first appearances of stone tools in the archaeological record. We now know that such evidence predates not just our species but also our genus (*Homo*). If stone tool creation does not arise with humans, it cannot be the first glimmering of human culture. Evidence of cave painting or other art, jewelry, or other modes of self-adornment, and particular patterns of hunting and complex food use have all been held to indicate that the bearers of human culture, traditionally called "modern" *Homo sapiens*, had emerged on the scene.

Most of these items begin to show up regularly at archaeological sites dating back 120,000 to 40,000 years ago. Assertions in the 1980s and '90s about genetic evidence of modern humans emerging around 120,000 to 160,000 years ago in East Africa (the "Eve" hypothesis) seemed to support the assumption that this was when human culture emerged. But in the past few decades it has become clear that humans who look more or less as we do were around by 200,000 to 300,000 years ago. The genetic evidence for the first appearance of contemporary humans has become less than clear. Recent work indicates that we may best think of ourselves as mixes of many previous human populations, rather than simply descended from one. We had no Adam and Eve but lots of great-great-great-great-grandparents.[31] All the evidence we have at present indicates that there was no clear line that, once crossed, gives us "modern" *Homo sapiens* and human culture. We did, in fact, evolve and are still doing so. Becoming human is ongoing.[32]

For example, while stone tool technology does not identify humans, particular types and uses of that technology do. By 1.2

million years ago, human ancestors were making stone tools far more complex than those of any other hominin species. By 500,000 years ago, ancestral humans were creating beautiful, highly symmetrical, stylized cutting and chopping tools that could only have emerged from multiple stages of planning, teaching, and artistry. For these tools, their aesthetic qualities often far surpassed any functional need.[33] These embellishments were created not because our ancestors needed them but because they wanted them, because they could create them, and they believed it was important to do so.

By 300,000 to 400,000 years ago, ancestral humans were more often than not using fire to keep warm, to modify tools and other items for their use, and to cook animals and plants, altering flavors and consistencies and nutritive values and changing forever what it means to eat. These patterns created a distinctively human behavior and experience, and all of it predates humans who look like you and me. By 100,000 to 200,000 years ago we see significant evidence that humans are exercising a distinctive creativity: making and using pigments to color their bodies and other objects, making and wearing beads, developing glues to bind things together, and creating new items that expanded their bodies' images and capacities substantially. Taken together, these elements indicate a human culture unlike that of any other organism. But the record clearly shows this culture growing and changing over time. There is no distinct point when our ancestors crossed over from nonhuman to human. Rather, we evolved in dynamic interface with cultural processes into the contemporary human niche.[34]

Various aspects of human culture emerged at different times in our history, coalescing into more or less the contemporary cluster over the past few hundred thousand years. But this process has by no means stopped. We keep evolving. To paraphrase anthropologist and philosopher Bruno Latour, we have never been modern humans. We are humans evolving, past, present, and future. Or to go back to Tim Ingold, we are constantly becoming human. It is just that, recently, we do it in more complex ways than in the past.

Over the past 400,000 years, various elements emerged that qualify as being central to *human culture*. Some died out almost immediately, staying only with the group that invented them. Other times these innovations spread across a few populations, like the use of carved ostrich eggshells for water vessels in Southern Africa. Occasionally, they spread across many groups, as with the development of carved figurines. As we come closer to our current time we see more and more evidence of the spread and maintenance of such patterns and behaviors. The appearance of human culture, just like human bodies, is evolutionary, emerging in fits and starts across great stretches of time and eventually coalescing into its contemporary forms.

At this point we have to engage the obvious question. What about language? We know that today, becoming human and human culture are inextricably bound to the capacity for language. The use of and immersion in language is a distinctively human reality. No other species displays this capacity. Some individual apes and parrots have been taught to use a few words, but no other species is inextricably bound, as the human species is, to language for everything its members do, think, and believe. Philosophers, psychologists, and linguists[35] have suggested that it's when we develop human language that human culture truly begins, that humanity becomes "rational." If that is true, how we believe is rooted in how we speak, think, and write. This is in large part accurate today, but less so in the past.

Language does not fossilize, but its development, the initiation of a human immersion in what the anthropologist Alan Barnard[36] calls the "signifying revolution," plays a central role in human evolution. So what can we actually say, with data, about language and culture in the deep past?

We know that the capacity to produce the range of sounds needed for human language is found in our closest living relatives, the living apes, and thus was present in our hominin ancestors, who also shared ancestry with the ape lineages. We have robust evidence that the ear canal of genus *Homo* had developed into the form that is most effective for hearing and isolating the sounds, pitch and frequency of human speech by at least 400,000 years ago, and that these particular morphologies are not found in the apes or in earlier human ancestors. So we can be fairly certain that human ancestors could both produce and effectively hear the sounds of speech, key tools of language, by 400,000 years ago.[37] Some researchers point to certain genetic complexes, such as the FOXP2 DNA sequences,[38] that seem to be shifting at around this time as additional signposts of the emergence of language. But the bulk of the evidence argues strongly against one or even a few "language" genes, or even a specific time stamp for the emergence of language. The physical structures, possibly the genetic infrastructures, and some of the material indications of language are clearly present by 300,000 to 400,000 years ago, but there is no definitive evidence of language use until much more recently.

We know that over the past 400,000 years the human material record grows more complex, and the majority of the items we presume to be evidence of human culture become increasingly prevalent.[39] I suggest that language is *becoming*, not appearing. There was no protolanguage that turned into language at a specific point. Instead, the capacity for language evolves in concert with the production and use of sounds, gestures, imagery, and the myriad processes in the human niche. Language must have evolved in diverse manners across many populations in concert with all the other processes of an evolving human niche, not as something separate from them.

Human culture and human language are covalent, intertwined in reciprocal causation and mutually restructuring. The dynamism of language expands the possibilities for human cultural processes, and both play a central role in humans' capacity for belief. The mutual entanglement of language, meaning-making, and the human niche of culture facilitates the presence of another human capacity that plays a central role in the development of belief: transcendence.

TRANSCENDENCE?

Transcendence, to paraphrase Immanuel Kant,[40] is the human capacity for recognizing that which is beyond the limits of all possible experience and knowledge (what I and many others refer to as a distinctively human imagination[41]). Anthropologist Maurice Bloch[42] connected this idea into the context of human evolution when he argued that we can see the critical transformation in the human lineage as it moves from engaging in transactional sociality (as most animals do) to adding a suite of transcendent relationships to its mode of social interactions. Humans become simultaneously transactional and transcendental beings.

Any account of human evolution must encompass the human capacity to imagine wholly novel, even transcendent realities. The specific understanding of how humans develop in and with the human niche requires both an explanation for how bodies, perception, and skills codevelop, mutually entangle, and influence their relationships and themselves, *and* how this process is related to our ability to imagine and, one might say, "experience" the transcendent.

The human capacity for transcendence has been implicated in philosophical and theological debates for millennia, from early Vedic texts and the Abhidhamma Pitaka of Theravada Buddhism, to Aristotle and Lao Tsu, from the precursors of the Abrahamic traditions to contemporary Jewish, Christian, and Islamic scholarship, and to a range of secular philosophers.[43] My contribution here is an attempt to integrate key elements of this discourse into human evolutionary history, with the goal of offering what I hope is an innovative synthesis. I want to emphasize that I am not rehashing the "human rationality" argument invoked by so many in the Western canon, from the Greek philosophers through Aquinas, Kant, Macintyre, and others.[44] I am not trying to create a deterministic or purely functional account of these capacities in humans. Rather, I am proposing that we consider the human experience as a dynamical system that has an evolutionary history and is still evolving, and that the capacity for belief emerges from these processes.

CHAPTER 6

How Culture Works

PSYCHOLOGIST Michael Tomasello once said, "Fish are born expecting water, and humans are born expecting culture."[1] Over countless eons, fish have evolved in an environment in which water was central to every aspect of their being, and they develop accordingly. A fish's body, its behavior, and its perception of the world are wholly inseparable from its existence in water. Over the last 2 million years, human development has evolved as a system that is inseparable from the linguistic, socially mediated, and constructed structures, institutions, and beliefs that make up the human niche. This human cultural niche[2] enables us to benefit from massive accumulation of information by countless individuals, groups, and populations; to be innovative in ways that no individual could be on one's own; and to learn and teach and socialize in ways unavailable to other organisms.[3] A distinctively human culture is part of our nature.[4]

Humans are not hardwired with any specific culture. We are "wired" to coacquire and develop such structures across our life span.[5] Rather than think of a human acquiring culture or developing a body, we can think of those aspects of the human as mutually entwined processes of becoming that unfold from the uterus to the grave. Three key aspects of these processes are the acquisition of skills, neuroendocrine development, and the capacity to create specific mental representations. These combine to give humans the capacity to think "off line" and experience a true imaginary, critical process for belief.

ENSKILLMENT, NEUROENDOCRINE DEVELOPMENT, AND MENTAL REPRESENTATION

Tim Ingold describes the development and refining of skills, which he terms "enskillment," as "the embodiment of capacities of awareness and response by environmentally situated agents." He elaborates:

> Neither innate nor acquired, skills are grown, incorporated into the human organism through practice and training in an environment. They are thus as much biological as cultural. To account for the generation of skills we have therefore to understand the dynamics of development. And this in turn calls for an ecological approach that situates practitioners in the context of an active engagement with the constituents of their surroundings.[6]

In short, the human niche is what humans grow into, with and against simultaneously. Enskillment is a physical, cognitive, social, and potentially transcendent process.

As we navigate our social and ecological landscapes, our bodies and minds acquire particular motor and social skills. All of this occurs in a particular cultural context. We embody the specifics of a language, particular modes of movement and behavior, appropriate mannerisms, and the use of and interface with clothing, food, and patterns of socializing, rules, laws, and customs. We also are immersed in the details of the beliefs of our home community and, if we travel, of the places and peoples we encounter. Throughout our lives, our minds and bodies are shaped by all of these experiences.

This honing of particular modes of perception and action creates the skills that humans acquire, use, and alter across their lifetimes. These skills include the use and manipulation of language, our mode of walking, the capacity to predict slight changes in the weather, the ability to read subtle social cues in faces and bodies, a "gut" sense of right and wrong, and the facility to deeply feel, physically and emotionally, a religious commitment or a political stance. These skills and others are not external features mapped onto the biological base of our bodies. They are materially real parts of us, embodied in our neurobiology.

Anthropologists Greg Downey and Daniel Lende, who offer an eloquent and accessible summary of these processes, point out that our neurobiology is always developing as we navigate the human niche and the dynamics of human culture. The skills we acquire are at once material, perceptual, and experiential. In their book *The Encultured Brain: An Introduction to Neuroanthropology*. Downey and Lende state,

> Neuroanatomy makes experience material—Neural systems adapt through long-term refinement and remodeling, which leads to learning, memory, maturation, and even trauma. Through systematic change in the nervous system, the human body learns to orchestrate itself as well as it eventually does. Cultural concepts and meanings become anatomy.... Rather than

one set of genes or an overarching system of meaning, humans' capacity for abstract thought emerges equally from social and individual sources, built of public symbol, evolutionary endowment, social scaffolding, and private neurological achievements.[7]

Our experiences, memories, and thoughts are biochemically embedded in our brains and bodies in a process that reshapes itself across our entire lives.[8] Our development in the human niche is literally coconstitutive with our neural and endocrine systems. This combination, the "neuroendocrine system," is the amalgam of bodily systems responsible—in us, as in all organisms—for maintaining homeostasis: regulating reproduction, metabolism, sleep, eating and drinking, energy utilization, and blood pressure. In many socially complex organisms, these neuroendocrine systems are often called "psychoneuroendocrine" because they are well understood to be in constant mutual interface with social behavior in social communities, and in humans with the experience of cultural institutions and processes.[9] Our bodies are literally cultural organisms. Cultural experiences are integrated into our skin, our muscles, our nerve fibers and neurons, and our entire physiological systems, and this process is central to the development, makeup, and function of our brain.[10]

One obvious way this plays out is how different people from different cultures interpret tastes of different foods, perceive "hot" and "cold" temperatures, and even on which aspects of a painting or a landscape their eyes first focus. Take two identical twins, separated at birth and raised apart, one in northern Alaska and the other near the equator in Brazil. As adults if you put them together and had them stand in a room at 60 degrees

Fahrenheit in shorts and a T-shirt, one would be freezing and the other warm. If you gave them raw salmon to eat, the smell and taste would stimulate a very different, visceral reaction from each. When you ask each to describe the same painting, they would likely point out different features in it as drawing their interest. The temperature, the salmon, and the painting are all constant; the experienced physical and perceptual reactions of each of the twins differed, despite having identical DNA, and very similar patterns of taste buds, muscles, nerve bundles, body hair, and rods and cones in their eyes. Each sees, tastes, and feels the stimuli differently because each has developed—become enskilled—in different cultural and ecological contexts. And this is not just for external physical sensations and interactions, it is in our cognitive processes as well.

The brain itself, as a hyperdynamic organ interconnected with all sensory and bodily apparatuses, shapes and is shaped by these relationships. We saw in chapter 3 that the expansion in the brain's neurobiological complexity was a central facet of our lineage's evolution and that it opened up possibilities for learning and innovation unavailable to other organisms. The human brain is only about 40 percent of its adult size at birth, the lowest known ratio among mammals, and it undergoes reorganization through much of our lives. That the majority of brain growth happens out in the world, rather than in the womb, makes the interplay between the constituents of the human niche and the development of the human body all the more critical. The brain and all its networks develop, lay down connections, and alter pathways and flows of biochemical signals in constant exchange with our bodies, our senses, the social and physical landscapes we inhabit, and our own thoughts, perceptions, and experiences of all these things.[11] This is the embodiment of culture.

Recent cognitive science argues that our cognition[12] is simultaneously enacted in our lives, extended beyond our brain, embedded in the social fabric in which we exist and are embodied, physically part of us, through the psychoneuroendocrine system.[13] Downey and Lende are speaking literally when they say that "cultural concepts and meanings become anatomy." As we develop in our niche of human culture, our experiences and perceptions shape and are shaped by our neuroendocrine systems, developing an anatomy and physiology that are part of the niche's cultural dynamic. That anatomy and physiology interact with the world to reconstruct the very niche that shaped it. It is by this process that humans develop skills throughout their lives.

It is not only physically that we interact with the world. One reason the human niche is particularly dynamic is our capacity to be cognitively unrestrained by our immediate material context. Like the fish swimming and becoming in water, humans are immersed in, and constitutive of, a semiotic[14] ecosystem that includes substantial symbolic components and a continuous potential for the imaginary and transcendent. Thus, unlike that of the fish, the human niche is shaped and bounded by things that are not always material or physiologically perceptible. Humans can see the world around them, imagine how it might be different, and translate those imaginings into reality—or try to. The human experience, and thus our skill set and our perceptual reality, is not limited to the tangible or material.

It is clear that complex and occasionally immaterial mental representations are a central process in human culture, and that the system underlying these capacities is especially salient in the contemporary functioning of the human niche. It is the dynamic combination of our psychoneuroendocrine processes, our capacities for and patterns of enskillment, and the complexities of the human cultural system that enable us to develop and deploy particularly complex mental imagery. This process gives us the mechanism to believe.

THE HUMAN MIND ENABLES BELIEF

The human capacity for belief is connected to the human mind. "Mind" is often seen as the cognitive processing of the brain, but it is much more than that.[15] Kevin Laland recently wrote that "no single prime mover is responsible for the evolution of the human mind. Instead [there are] accelerating cycles of evolutionary feedback, . . . an irresistible runaway dynamic that engineered the mind's breathtaking computational power."[16] He argues that the most critical elements of human cognition[17] cannot effectively be understood in isolation. Rather, it is the mutual co-shaping of each by the others that is the critical unit of analysis.

Human minds are not best understood as the product of changes in external environments. They do not simply result from particular patterns of natural selection that favored specific cognitive adaptations. Instead, Laland suggests, our mental abilities evolved via a "convoluted, reciprocally caused process, whereby our ancestors constantly constructed aspects of their physical and social environments that fed back to impose selection of their bodies and minds, in endless cycles."[18]

The "mind" is the sum of the elements we have been discussing throughout this book: it is the set of skills and processes that enable humans to think and act as they do. It is the amalgam that Downey and Lende are describing when they write, "Cultural concepts and meanings become anatomy."[19] The mind is the product and process of human culture, and the central outcome of evolutionary dynamics in the human niche.[20]

For our inquiry here, we can focus on the key capacities of the mind that undergird and enable our capacity for belief. These are the distinctive human abilities of imagination and complex mental representation.

Philosopher Anna Abraham[21] suggests that our ability to imagine can be seen in our capacity to conjure up images, ideas, impressions, and intentions. She argues that the conceptual space the human imagination spans is "stupendously vast, stretching across the real and the unreal, the possible and the impossible." The imagination is at once "spontaneous and deliberate, ordinary and extraordinary, conscious and unconscious, deriving from the outer world external to our bodies as well as our inner world." Abraham, quoting Nigel J. T. Thomas, emphasizes that "imagination makes possible all our thinking about what is, what has been, and, perhaps most important, what might be."

Moving beyond simple dictionary definitions of imagination (such as "the mind's creativity and resourcefulness"), Abraham lays out a model of the human imagination as having five modes of operation: mental imagery (perceptual and motor), phenomenological (emotional), intentionality (recollective), novel combinatorial (generative), and altered states. Thinking of imagination as active in these arenas enables us to connect it to specific cognitive, behavioral, and material processes directly associated with the act of belief. When we believe, we use mental representations to form basal understandings and heuristics for explanation. We develop phenomenological (emotional) investments in these images and concepts and are able to retrieve and manipulate their meanings and share them with others. By placing all of these processes into interaction, we generate new meanings and new understandings.

A mental representation is an internal cognitive sign that

represents an external condition. The condition need not be something actually experienced: the mind may use symbolic representation to create a sensation of things that are not now, and perhaps have never been, available to the senses. For our discussion about the mind and human belief, we are particularly interested in the process of developing mental imagery of things that are not present, and especially those that have never been experienced.

Philosopher Peter Gärdenfors and colleagues[22] use the labels "cued" and "detached" to describe mental representations. "Cued representation" refers to the mental representation of things that have been experienced in the past and are cued into "re-experience" by something in the subject's current external situation. For example, if I enjoyed an apple a few days ago and then today walked by the grocer's and saw fruit in the window, I could recall the apple I ate and the experience of consuming it. The representation is cued by the experience of seeing the apple. The critical capacity here is that I can actually "see" the apple in my mind and perhaps also "taste" it and possibly "hear" the crunch. All of these experiences are active events in my mind—they have biochemical neurological processes associated with them—that can cue a wide range of related memories, experiences, thoughts, and ideas.

"Detached representations," on the other hand, are mental representations of objects or events that are not present in the subject's current or recent external context and thus could not directly trigger the representation. A memory evoked independent of context is one example. Such memories need not be based on something that actually occurs in the external world. They can be imaginary. I could be daydreaming on a train and conjure a winged octopus with twelve arms, or maybe an image of an animal that bears no resemblance to any I consciously remember. In either case the mental images are largely detached from my material experiences at the moment, and yet they are accompanied by all the same bodily, sensory, and cognitive processes that characterize cued representations. Both types of representations are real in my experience, neurobiologically and perceptually, but the detached representation is especially powerful in the context of belief.

In order to have such detached representations there must be (a) a substantial capacity for imagination (especially in the generative sense noted by Abraham) and (b) an ability to block out, at least partially, one's current environment. This is a tall order for any organism, given the deep evolutionary importance of remaining aware of your surroundings (to guard against things trying eat you, among other dangers). Such a capacity is centered in the brain's frontal lobes, the very location that, in the human lineage, became exceptionally more elaborate during the last few million years.

In humans the frontal lobes constitute approximately 30 percent of the volume of the cerebral cortex—more than in any other mammal—making this one of the few locations in the human brain that differentially expanded rather than simply scaling up in size.[23] The frontal lobes are critical in emotional, social, motivational, and perceptual processes in all mammals, and are particularly important in coordinating decision making, attention, and working memory. Intriguingly, another critical area for complex function, the cerebellum, has also expanded in the great apes (relative to other primates), and even more so in humans[24] The human cerebellum contains four times as many neurons as the neocortex, and it has expanded at several times the rate of the neocortex in apes. It is estimated that, over its history, the hominin lineage added approximately 16 billion more cerebellar neurons than expected for our brain size.[25] The cerebellum's key functions include social sensory-motor skills, aspects of imitation, and production of complex sequences of behavior. In short, while certain capacities for detached mental representation may lie in the primate lineage, especially in apes, it appears that the human brain underwent a specific structural transformation that made detached mental representation more effective and more expansive.

Gärdenfors asserts that this capacity for detached self-awareness, where an agent has self-representations that concern her future or past conditions and can imagine herself as having different properties in different domains, paves the way for "the self-ascription of properties that the agent does not in fact have, and thus for counterfactual reasoning about the self." These combined capacities enable a person to think in the here and now *and* to project herself into future or alternative contexts where the external condition or the vision of self may be totally changed. This is an absolutely critical skill when we draw on our range of cognitive and social resources, our histories and experiences, and combine them with our imagination to "see" and feel and know "something." In short, when we believe.

Philosopher of biology Kim Sterelny[26] and neuropsychologist Merlin Donald[27] call this capability "decoupled representation." Michael Tomasello[28] and his colleagues describe this capacity as a critical differentiating factor in the divergent abilities of young humans compared with chimpanzees. Human children are able to innovate and collaborate to solve tasks, and they think more imaginatively and collaboratively. This entanglement of imaginative abilities with the capacity for detached representation offers a structural mechanism for the act of believing.

If we consider this aptitude, and all that we've covered about human neurobiological development and actions, in light of the diversity of geographies, ecologies, histories, and cultural experiences in human communities today, we can see why there is so much variation in *what* we believe.

Research by cognitive scientist Shihui Han, and many others, reveals that the brains of people from different cultures respond differently to specific stimuli and social contexts.[29] Han found that adults from China, Japan, and Korea differed from adults from Western Europe in brain imaging responses to face-recognition tests, perceptions of others' emotions, empathy for others, some semantic relationship tests, and even patterns of self-reflection. Such work also reveals across many societies that cultural knowledge systems affect brain development and activity in pain sensing, modes of empathy, monetary reward tests, and how we process social feedback about our actions. Not surprisingly, anthropologist Tanya Luhrmann and colleagues[30] are able to demonstrate that people's perceptions of themselves and others, and their perceptions of transcendental experiences, are tied to their culture's representations of the world around them. This variation specifically reflects how culture, brains, and bodies work with and on one another and how this maps to the specifics of how human societies, histories, and daily lives vary from place to place. Where we grow up and who we grow up with affects the details and structures of our belief.

How Human Culture Makes Belief Possible

The human body and mind evolved as a system in which physiology and neurobiology are mutually coconstitutive of the linguistic, socially mediated, and constructed structures, institutions, and beliefs that make up the human niche. This process is human culture. Let me summarize the details:

- Neural and endocrine systems develop as humans learn to orchestrate themselves within a cultural context (our niche). Through this process, social concepts and meanings become anatomy, which in turn interfaces with and potentially reshapes the broader cultural milieu from which social concepts and meanings emerge.

- The specific patterns of how we use our bodies, minds, and skills are grown and incorporated into the human organism through practice and training in a given environment. They are thus simultaneously biological and cultural. Skills are contingent on the capacities and constraints of our bodies and our relationships with one another *and* the social, historical, and material environments we inhabit.

- The particular evolutionary histories and processes in the human niche have led to distinctive neurobiological development and cognitive expansion of the human brain. In concert with the processes and structures of human culture, this expansion has enabled humans to develop the capacity for extensive detached mental representations and a particularly powerful capacity for imagining.

- The shape and boundaries of the human niche are not always material or circumscribed by cued representation. We are thus open to influence from transcendent experiences in addition to those that are specifically cued or materially experienced.

- This enhanced capacity for detached representation and imagination, and the complexity and diversity of our

social and ecological milieus, enable humans to experience, create, and develop highly diverse skills in perceptions and awareness that are not contingent on material
reality. These perceptions may include transcendent
experiences.

These five elements are the basis of the human capacity to
imagine, to be creative, to hope and dream, to infuse the world
with meanings, and to cast our aspiration far and wide, limited
neither by personal experience nor material reality. The patterns
and processes I've outlined in this and the previous chapter constitute the basis for *how* we believe.

To believe is to make a commitment. The understanding of
human culture I've outlined here, as a dynamic system in concert
with our psychoneuroendocrine processes, draws connections
between human evolutionary histories, our neuroendocrine systems, and the patterns and process in the human niche that enable
our capacity for belief. *How* we believe is specifically tied to distinctive cultural processes of becoming that emerged across our
evolutionary history. This account offers us two critical insights
for understanding why belief is so central to human existence.

The first is an anthropological dictum: "cultural constructs
are real for those who hold them." Many people, on hearing that
something is a cultural construct,[31] often infer that it is therefore
not "real" but only an imaginary or socially agreed-upon set of
assumptions. It is "all in your mind." This is wholly incorrect. For
humans, there is no such thing as "all in your mind," because the
mind is not just in our heads and doesn't comprise only imaginings or agreed-upon external symbols. Our heads are part of our
bodies. All perception and functioning of the neuroendocrine
system is shaped and structured by internal and external inputs,

which in turn are filtered and molded by one's experiences, skills, and beliefs.

A cultural construct is a specific conviction that is widely shared among members of a culturally defined group. Thus, it is part of the group's perceptual and physiological experience, as well as constitutive of the social ecology in which the group exists. Often cultural constructs are, in fact, beliefs. Remember, to believe is to invest in something, utterly, wholly, and authentically, such that it is one's reality. So cultural constructs *are real* for those who hold them. That is the way the human mind works.[32]

Thus the second insight: for humans, what *is* and what *should be*, two central components of every human belief system, are deeply contingent on where and how we develop.

By this point I hope it's clear that a deeper understanding of what human culture is offers insight into our capacities for belief. The integration of cultural processes, evolutionary histories, and how our bodies and brains develop in interface with our niche provides the core understanding of how we believe.

To these insights on culture and belief I add a coda. As the famous semiotician Charles Sanders Peirce put it, "All the greatest achievements of mind have been beyond the power of unaided individuals."[33] That is, we seldom develop beliefs alone. Culture is a communal affair, and how and what we believe is not just connected to the communities in which we participate. It is shaped and facilitated by our extensive and obligate sociality. What we call a "belief system" is a particular pattern of collaborative imagination and commitment by a social group. Among the most common and powerful of these are three that we are now in a position to discuss: one we generally call "religion," another we term "economies," and the final we call "love."

Part 3
Religion, Economies, Love, and Our Future

CHAPTER 7

Why We Believe versus
What We Believe

THAT HUMANS believe is a distinctive, powerful reality that has shaped our planet. Understanding both *why* we can and do believe and *what* we believe is critically important to our lives and future. While the answers to *why* and *what* we believe overlap, they are not the same. Both matter, but in different ways.

So far in this book we have focused on the evolution of our capacity for belief and how the human niche and human culture facilitate the production of belief. Now we turn to specific beliefs. The first three chapters in this section delve into three patterns of belief—religion, economies, and love—to examine the interaction of *why* and *what* we believe. In the fourth chapter we discuss how belief can matter, for worse and for better.

I want to begin by offering a very brief recap of what we know about *why* and *how* we believe.

The first and most obvious answer to the question "Why do we believe?" is that we believe because we are human. Just like our large and complex brains, our ability to walk on our hind legs, our nimble fingers and hands, and our ability to make tools, the capacity for belief is part of our distinctive evolutionary history. To be human is to be able to believe.

The core processes that facilitate our capacity to believe evolved in genus *Homo* over the last 2 million years, during which our cognition and behavior were shaped by the feedback processes in toolmaking, foraging, cooperative creativity, and communal caretaking of our young. *Homo* perceptual landscapes shifted as their childhood development lengthened, and as they controlled fire, created meaning-laden materials, and expanded around the planet. The feedback among all of these processes and evolving bodies and minds resulted in the emergence of the necessary infrastructure for the human *capacity for belief.*

Material, social, and ecological complexity ratcheted up as populations of *Homo* developed more dynamic interactions between themselves and the world over the last 300 to 400 millennia. This process created the critical cognitive and demographic infrastructure that facilitated the connections and exchanges between populations necessary for the development of environments and societies that enabled shared beliefs, and eventually belief systems, to emerge.

In the last few hundred thousand years, meaning-making in human populations grew and eventually became universal. The human niche became infused with a capacity for innovation, imagination, and even transcendence. Humans went from being socially complex transactional beings to groups of organisms who exist simultaneously in both transactional and transcendent realities, and who use imagination and belief to reshape themselves and the world around them.

Over the last 15,000 years the human niche changed and diversified still more with the development and expansion of domestication. Sedentism; communal architecture; larger and larger communities; changing notions of identity, memory, and politics; and the creation and management of substantive storage systems reconfigured human existence. With the new landscapes came new concepts: property, increasingly structured and diverse types of inequality, and large-scale, impersonal violence and warfare. The basic structure of the contemporary human world emerged.

HOW WE BELIEVE

The neurobiological part of the answer to *how we believe* is twofold. First, over evolutionary time, our brains underwent substantive change in the frontal lobes and cerebellum, and across the neocortex. Compared to those of other primates, human brains develop more slowly, fold more intensely, build more connections, and have greater plasticity in function. Second, because more of this development happens outside the womb, our brains are extensively coconstructed through interaction with our increasingly complex social and ecological landscapes as we grow up in the human cultural niche. As Downey and Lende write, "Neuroanatomy makes experience material. . . . Cultural concepts and meanings become anatomy." The languages we speak, the foods we grow up with, the political, economic, and gender systems we participate in—all infuse themselves into our neurobiology, shaping how we perceive and interact with the world.

But we are more than our brains, and the capacity for belief is not rooted simply in neurobiology. Our brains do nothing without our bodies. Our bodies are never outside of our social-ecological contexts, and these contexts are enmeshed in a linguistically and behaviorally mediated human cultural niche that is built from the actions, perceptions, experiences, and institutions of people past and present. Through the processes and structures of human culture, humans develop the capacity for extensive detached mental representations and a particularly powerful ability to imagine.

These enhanced capacities and the complexity and diversity of our social and ecological lives enable us to experience, create, and even develop diverse skills that are not contingent on material reality. These skills can develop in ways that incorporate transcendent experiences as part of the human tool kit. Together these elements enable humanity to draw on a range of cognitive and social resources, histories, and experiences; combine them with our imagination; and think beyond the here and now when developing our mental representations in order to "see" and feel and know "something." To believe.

Let me reiterate what I said in chapter 3: the capacity for belief is not simply an emergent property of being human. It is not some ephemeral thing floating above the material reality of the human experience, nor is it an add-on or by-product of human evolution. The ability to believe is central and material to the human system. It is part of us in the same way that fingers are part of our arms and hands.

Fingers are core aspects of the human organism that have been modified over evolutionary time to dramatically expand our options for interacting with the world and each other. Our hands and arms have been shaped over evolutionary time so that their terminal ends contain structures and capacities that enable us to do much more than we could without such components. Today we use fingers to send texts, make craft cocktails, build houses, paint pictures, and touch those we care for. Our capacity for belief, like our fingers and hands, is a core aspect of the human organism, critical to the human ability to engage with and shape the world—

to become in the human niche. With belief, however, we extend our reach much farther and more powerfully than we ever could with hands and fingers.

WHAT WE BELIEVE

Understanding the why and how of belief in general is not enough. Why do we believe *specific* things? Why do those beliefs bring together, break apart, and reshape communities of humans? Most of us, in our everyday lives, are not confronted with big, overarching evolutionary questions. Usually we don't wonder, *Why do we have the ability to believe?* Instead we ask, *Why do we believe the things we do?*

The specifics of a religious faith, the details of gender roles and relations, the reasons for hate—the "what" in beliefs—matter deeply. The combination of patterns, contents, and contexts of specific beliefs, not just the capacity for belief, has the most significant impact on human lives. We call these combinations "belief systems," and in the details and histories of these systems we find some of the answers to why we believe what we do.

Effectively engaging *what* we believe relies on our ability to place our understanding of the capacity for belief in dialogue with the specific contents of particular belief systems. All belief systems emerge over time, and usually they represent a conglomeration of features of the human mind, societies, and histories.

In the next three chapters of the book I examine specific cases of what humans believe, in an effort to show how each is a powerful and important blend of our past and present, our capacities and proclivities. These examples reflect our tendency to draw on our range of cognitive and social resources, our histories and experiences, and combine them with our imagination to develop mental representations, and to utterly invest in these representations so that they become our reality. Religion, economies, and love are three of the most powerful realities for humans worldwide.

CHAPTER 8

Religion

RELIGION as we know it is very recent—and we are not. Human culture and our immersion in belief enabled our capacity to be religious to emerge over evolutionary history. But the development of religion as a fixture of contemporary human identity was facilitated by very recent events. In this chapter I offer an evolutionary explanation for why so many humans believe in religion.

Formal, large-scale religions and religious institutions—such as Christianity, Hinduism, Buddhism, and Islam—have roots going back no more than 4,000 to 8,000 years.[1] The distinctively human lineage is more than 2 million years old, and for nearly 90 percent of that history we have scant material evidence that anything like transcendent experience or an acknowledgment of the supernatural, both critical to religion, was prominent in our ancestors' lives.[2] Yet today, perceptions of transcendence and the supernatural are nearly universal across humanity, and a majority of humans describe themselves as belonging to a religion. Here I lean again on Kant for the term "transcendence" to describe the human capacity for recognizing that which is beyond the limits of any possible experience or knowledge. I, like many others in the anthropological and psychological literatures, refer to such a capacity as a distinctly human imagination but do not tie it necessarily to religion. However, for many theologians and philosophers, and most people on the planet, our capacity for transcendent experiences is most often associated with religious belief. These beliefs, however, do not stand alone; they are interwoven with institutions, traditions, rules, practices, and histories ("religions") that shape the everyday experiences of billions of humans. For most humans, the question "Why do we believe?" is deeply connected to religious contexts.

Today nearly 6 billion people, about 83 percent of the world's population, explicitly identify themselves as religiously affiliated. Religious experience of one sort or another is a daily activity for most people, and religion as an institution is woven into the hearts of the societies and nations in which we all reside, whether we personally are "religious" or not. Today the capacity to be religious is found in all of humanity, and therefore it must have some evolutionary relevance and history.[3]

Can being religious be understood as a central feature in human evolution?

This question is particularly tantalizing to those scientists, like me, who are interested in excavating and reassembling the paths by which humans arrived at the significant ubiquity of religious beliefs today. It is also interesting for the many people who experience elements of transcendence but do not accept any contemporary religious doctrine, as well as for religious individuals who understand that our evolutionary histories are relevant to a deeper understanding of our lives and want to incorporate evolutionary data and interpretations into theological and philosophical views of religion. In short, there is much interest in the evolutionary roots of religious belief.

Religion and Human Evolution

Anthropologist Roy Rappaport[4] suggests that "in the absence of what we, in a common sense way, call religion, humanity could not have emerged from its pre- or proto-human condition." Sociologist Robert Bellah,[5] following psychologist Merlin Donald,[6] agrees and argues that religion emerges from, and nurtures, the very processes of meaning-making that are central to the human niche.

While I do not think that "religion" made us human, it certainly has massive impacts on the processes and experiences of humanity, and thus is central to an understanding of becoming and being human.[7] I propose that multiple lines of evidence suggest that *having religious beliefs* is much older than formal religious systems, structures, and institutions. And this pattern offers us particularly important insights into human belief. To be clear, I do not intend to question the truth claims of any specific religious tradition. Nor am I trying to identify some "ur-religion" that gave birth to all of the contemporary varieties.

The assertion that having "religious beliefs" is much older than formal religious systems raises problems for someone who is dogmatic about the truth claims of one particular religious system or faith practice relative to others. I contend, however, that what I present does not conflict with most of the theological precepts of most religious traditions. Note that I said, "most." I unavoidably raise some insurmountable incompatibilities with threads of certain theologies that directly ignore the available data from the fossil, archaeological, and geological records.

Having said this, for the rest of this chapter I keep one foot grounded in the material record of our evolutionary past. We gallop through some data and look at a range of evolutionarily influenced ideas about why our capacity for belief has generated a specific kind of belief, and why this in turn has facilitated the emergence of religions and religious institutions. Be warned: I do not satisfy anyone who is hoping for an evolutionary explanation of any specific theology or religious tradition, as I think such endeavors are folly.

CLARIFICATIONS

First, and critically, I need to clarify the conceptual difference between "religious" and "religion."

I use the term "religious," borrowing from Clifford Geertz,[8] to mean *the use of one's capacity for belief in the context of becoming with particular perceptual, experiential, and agential practices, involving the transcendent, that act to establish powerful, persuasive, and long-lasting moods and motivations that may be but are not necessarily tied to specific formal doctrines, practices, texts, or institutions.*

"Religion," borrowing from Emile Durkheim,[9] is *the formal coalition of religious beliefs and practices (rituals) and the material symbols and structured institutions that unite them into a single community via specific theological doctrine and ritual.* As philosopher Tim Crane[10] puts it, religion reflects not just the having of transcendent experiences but the *systematic and practical attempt by human beings to find meaning in the world and their place in it, in terms of their relationship to something transcendent.*

Many theologians, philosophers, and others may balk at this division between "religious" and "religion," and it can certainly be challenged. But please bear with me as I try to demonstrate that this method is necessary for an effective evolutionary analysis of this topic.

A critical reason for this division is the relative youth of what are termed "formal" or "world religions," such as Christianity, Islam, Judaism, Buddhism, and Hinduism. Any attempt to describe patterns and processes over the longue durée of human evolution demands that we portion off large-scale contemporary religions, given that the institutions, texts, ritual structures, and much of the content of their doctrines represent particular historical patterns that have no materially identifiable roots or analogs in the human record deeper than the past 8,000 years or so. Contemporary religions as institutions are just that: contemporary.

Most evolutionary approaches to human religious belief present it as either a straightforward cognitive adaptation, or else as a "spandrel," meaning a by-product, not necessarily helpful to survival, of unrelated evolutionary developments. The general conclusion is either that religious belief serves a specific evolutionary purpose, or that our evolved cognitive capacities led to psychological systems that make us prone to developing supernatural explanations for phenomena and experiences.

Some evolutionary hypotheses claim to offer explanations for key contents of religious belief systems. I believe they do not succeed in these efforts. Such approaches offer functional and often reductive narratives about specific transcendent experiences— but these experiences, almost by definition, resist reductive, materialistic descriptions. I do not think an explanation of the theological and practical aspects of immersion in a religious tradition is obtainable under a singularly evolutionary argument.

But neither am I a Cartesian dualist.[11] I do not accept a ghost-in-the-machine[12] explanation that places mind or "soul" at the center of the human experience of belief. Pseudo-Cartesian approaches offer no more sufficient answer to why religious belief emerged than accounts based on adaptation or spandrels. My experiences as an anthropologist and zoologist force me to view all such reductive or dualistic explanations as either incorrect or incomplete. They are certainly unsatisfactory in the light of what we know about human evolution, neurobiology, physiology, history, culture, and religious experience. Even without belonging to a "religion," one can be "religious" by having beliefs that engage transcendent experience, and possibly the supernatural, but that do not connect or derive from a specific doctrine or coalition of formalized or institutionalized beliefs, symbols, and practices. Such a capacity needs explanation, and the answer is inevitably not simple.

Finally, we need to acknowledge that any given religion unites particular beliefs, doctrines, and rituals with material symbols and structured institutions in a way that has meaning for a particular community. Without access to the experience and cultural realities of the people who participate in the religion, we cannot, using only the material in the archaeological record, identify it with any confidence as a "religion." Even what we might term small-scale, "traditional," or animistic religions are not directly discernible from material evidence in the deep past. In this chapter I do not make any attempt to identify past "religions" and instead lay out a material case for why we believe and why that belief often appears to take on religious aspects.

A Brief History of the Evidence for Meaning-Making in Human Evolution

Meaning-making is the capacity to think anew and create material realities out of these new thoughts. Indications of meaning-making in the human past offer material evidence for when and how humans may have developed the capacity for transcendent experiences that establish powerful, persuasive, and long-lasting moods and motivations—the baseline for religious experience.

Anyone walking into the Paleolithic art-filled caves at Altamira, Lascaux, or Leang Timpuseng experiences an aesthetic draw, a sort of transcendent shiver, without even knowing anything about the images or the culture that painted them tens of thousands of years ago. A similar experience happens when entering the cathedral in Toledo, the Blue Mosque in Istanbul, or the temple ruins at Borobudur and Prambanan in Java.

In the latter cases we either are immersed in the culture and beliefs of the creators of these places or have some knowledge of them, but in the former cases we have no such knowledge. We cannot know what meaning the images on the cave walls held for the humans who painted them so long ago. Yet the very fact that we're human enables us to experience some commonality of sensation and perception. We can intuit meaningful imagery and structures in these places and connect across space, time, and culture. It is the human capacity to believe, perhaps the closest we can ever get to time travel, that enables us to respond to the material representations of the beliefs of other humans.

Cave paintings, temples, and monumental architecture offer concrete evidence for human belief systems, but most of these magnificent constructions are very recent. Before about 70,000 years ago we find no cave art, and not until the last 30,000 years or so does such art become extensive and commonplace. The building of identifiable temples, gathering places, and monuments reflecting belief systems is even more recent, appearing only in the last 10,000 years.[13]

One might argue that such edifices and art suggest that the capacity for belief systems emerged at one of these points, from either a cognitive/neurobiological shift or some form of spiritual or divine revelation. As an evolutionary scientist, however, I know that a single shift that changed everything, whether evolutionary or revelatory, whether 70,000 or 30,000 or 10,000 or even 2,000 years ago, is an impossibility. That is not the way the processes of evolution work, nor is it what is reflected in the fossil, genetic, or material record of the past. The available data refute such an assertion. It is not that major shifts, changes, and innovations did not occur at these points, but those changes were necessarily rooted in the bodies, behaviors, and minds that came before them. Belief systems did not appear de novo.

Evolutionary questions about belief and belief systems have to involve data, and the key data for past behavior comes to us as material remains. So the first question to ask about the capacity to be religious is: is there material evidence for anything like symbolic representation or structured systems of meaning-making, religious or otherwise, in the deep past?

The answer is both yes and no.

Let me clarify the no answer first. Material evidence of meaning-making could be found only in the items earlier humans made that are preserved. If we are specifically looking for material symbols that reflect belief and belief systems of the deep past, we are out of luck. A symbol is something that stands for something else, based on the mutual agreement of the community using it. That is, the symbol's meaning is based on convention, not on any similarity or contiguity between the symbolic item and what it's supposed to represent. For example, a stop sign, red light, and the word "stop" are all symbols that we have mutually agreed on to represent the meaning of "to halt movement or discontinue action." There is no way that any outsider to this agreement can know that, unless informed by an insider. Therefore, without access to the actual context of the meaning-makers, we cannot accurately know what a symbol means.

Figure 6. A female Paleolithic figurine, Venus of Willendorf. Credit: Wellcome Collection. CC by 4.0.

When we go into the material record of the human past and observe an item many think of as a "symbol," such as the figurine known as the Venus of Willendorf, we have no way of knowing what it meant, or even if it was a symbol at all. We cannot truly know the meaning of the Venus of Willendorf for the people who made her. Still, it's clear this item was modified by humans to reflect certain shapes that are identifiable to contemporary humans as bearing meaning, and thus were likely to also represent certain ideas or meanings to the people who made it. The figurine is material evidence of meaning-making. If we drop the "symbol" label, we can then use the figurine to help us think through the evolution of belief systems.

Biological anthropologist Marc Kissel and I recently proposed a way around the reliance on "symbol" in thinking about meaning-making in the past.[14] To do so we drew on the semiotics theory of philosopher Charles Sanders Peirce.[15] Peirce is most widely known for his development of the sign trichotomy of *symbol*, *icon*, and *index*. The symbol we've already defined. An *icon* is a sign that bears a semblance or likeness to what it represents (like the image of a bike to denote a bike lockup), and an *index* contains a factual connection to what it represents (like dark storm clouds portending rain). But Peirce also proposed another trichotomy of signs that does not rely on symbol and the assumptions associated with it.

For this other trichotomy Pierce proposed *qualsign*, *sinsign*, and *legisign*. A *qualsign* signifies something through a quality it has. For example, in a blue cloth, "blueness" is a qualsign. A *sinsign* is one that uses essential facts to convey meaning. When a weather vane shows which way the wind is blowing, it is a sinsign. Finally there is the *legisign*, which is a sign vehicle based on convention. We can see that the sign has a specific meaning because it shows up in multiple places and evokes the same perception response. That is, if there are multiple examples of the same type of human-created material item, and each conveys or contains or evokes similar sensations, then we can say it reflects a convention among the group making the items. They are intentionally replicating a material item with similar characteristics— we assume with the same intended impact, but we cannot prove that, or know why that impact was desired. In other words, it may be a symbol, but we cannot know that. We can, however, assert that the legisign meant something to those who made it, as is evident from the repeated creation of items evoking specific sensory responses.

Figure 7. Multiple Venus figurines.
Source: The Natural History Museum / Alamy Stock Photo.

Thinking this way we can go back to the Venus figurine and see that from between 18,000 and 30,000 years ago across much of western, southern, and eastern Europe, several remarkably similar figurines have been found.[16] They are not identical, but share many features in shape, texture, size, and style of creation. They are replicas of a legisign. They offer an indication that multiple groups of people were creating material objects that represented a set of shared sensations or mutually understood meaning. The presence of legisigns offers evidence of meaning-making, whether symbolic or not.

Identifying objects as replicas of legisigns is not as exciting as thinking that they were fertility symbols, images of leaders, or reflections of sexuality. But we have absolutely no way of knowing what these figurines actually meant. We can show that they were created in the same ways and infer that the peoples who made them undoubtedly had beliefs about them that were shared. We can safely argue that the recurring presence of the figurines reflects at least the possibility of a belief system. But we cannot identify the contents of that system.

To investigate the origins of belief systems that involve meaning-making (a necessary precursor to religion) we have to ask: how far back do we have evidence of legisigns indicating the possibility of shared systems of meaning-making and thus particular patterns of belief?

A caveat: one might assert here that the concept of the legisign also applies to stone tools. And one might be right. We know that complex and aesthetically beautiful stone tools emerge in the archaeological record by at least 1.5 million years ago, and by 500,000 years ago are often shaped to be much more symmetrical and aesthetically pleasing than is needed for the job the tool was produced to do. But because we are thinking about meaning-making, belief, and the possibility of transcendent experience, we should stay away from items whose material function is identifiable from detailed analyses. Tools, even if they also have a meaning-laden, possibly transcendent aspect, always remain rooted in the here and now through their explicit tie to a material and practical use. Therefore, let us ignore tools here.[17]

The earliest possible examples of meaning-making may not be legisigns: we have not found any temporally concurrent replicas of them. But they are interesting. Researchers found a bovine shinbone, dating to around a million years ago, that appears to have a series of parallel engravings on it. Two suggestive pieces of yellow ochre[18] are also found at a very early site. Both pieces of ochre show evidence that they were modified to extract powder, possibly for pigment making. More recently, researchers found a clamshell on whose surface is carved a zigzag image, almost a "doodle," dating to about 400,000 years ago. Around this same time we find the Berekhat Ram and Tan-Tan figurines,[19] which are naturally human-shaped stones that were modified to look even more human. Between 200,000 and 300,000 years ago we find other examples of bones with etchings, evidence of ochres being used, and evidence that small stones and shells have been modified for use as beads.[20]

Early examples of meaning-making in *Homo* groups are scarce. In our overview of the currently available data Marc Kissel and I called them "glimmerings," rare and potentially isolated occurrences that demonstrate that early humans had *the capacity to create items that contained a particular pattern of meaning* but that the context for shared and sustained meaning-making was not yet present. But it was just around the corner.[21]

At this point, it's worth taking a moment to reflect on recent evidence for possible burials, as intentional burial is almost certainly associated with meaning and belief. Two examples are more than 200,000 years old.

The first dates to approximately 400,000 years ago, at the site of Atapuerca at a location called Sima de Los Huesos (pit of bones). It is an opening in a cave with a long drop into a pit, with the remains of some 28 individuals of genus *Homo* at the bottom.

No explanation for how the bodies got there, other than that they were purposely dropped into the pit, is supported by any evidence. There is one stone tool among them, a beautiful stone hand ax, exquisitely carved from a multicolored rock that came from many kilometers away. The hand ax was never used, merely made and tossed into the pit with the bodies.[22] That is all we know.

The second example comes from a recent discovery in South Africa, the Dinaledi chamber. Between 236,000 and 335,000 years ago, a group of small and slight members of genus *Homo* called *Homo naledi* carried their dead deep into an underground chamber and deposited them there.[23] The chamber in which the remains were found is extremely difficult to get to, more than 100 feet underground and requiring crawling, climbing, and squeezing though very small passages. They also had to drag the dead with them. And yet these members of genus *Homo* risked much to venture into the darkness and place these remains in a small chamber deep inside the cave.

We do not know why the people at Atapuerca and Dinaledi placed their dead in caves. We don't know if these populations were directly on the lineage to contemporary humans—probably not. But they inhabited the human niche in the exciting period between 100,000 and 400,000 years ago, when *Homo* lifeways were showing increasing evidence of meaning-making and contemporary humanity was developing.[24] The inner ear and vocal apparatus for language had developed by this time, and the neurobiology for speech was likely in place for many populations. We see evidence of a substantial uptick in the complexity of *Homo* populations' lifeways across Africa and Eurasia. Fire use becomes ubiquitous. There is evidence for the creation of glues and pigments and the use of more complex manufactured tools.[25]

We have already seen that the human ability to imagine, hope, and believe was ratcheting up. The increasingly complex material items, and the behaviors associated with making and using them, changed the landscape of human sensation and perception. Many of the items reflect the translation of inner perceptions onto external items. They are novel imaginings made material, influenced by the material world but not constrained by it.

The human niche was changing. During this period people were exchanging more complex information, and manipulating more types of materials and creating more uses for them. Stones, bone, ochres, beads, and bodies were finding their ways between and among populations of the genus *Homo*. The ways in which human communities interacted with the world and each other were deepening in complexity, a process in which the capacity for meaning-making played a role.

Between 80,000 and 200,000 years ago, it is extremely likely that language and genus *Homo*'s developing capacity for detached mental representation fueled creativity and innovation in all aspects of their lives. We have evidence of legisigns from this period: carved ostrich eggshells, beads, ochre and pigments used to paint on bodies and tools, and many more examples. By 70,000 years ago we see cave art; by 40,000 years ago, figurines. The human niche was growing replete with meaning-laden objects.[26]

We see evidence of increased interactions between groups and of social and material connections across wider geographies.[27] Augmented connections, materials, and communication create a virtuous cycle, nourishing the fire of imagination and opening humans' minds to a world that is more than merely material, more than the here and now, in which novel ideas can become material reality.

The evidence strongly suggests that it's in this time period that humans develop more linguistic and systematic modes of sharing beliefs.

But Is Any of This Religion?

We can certainly say that by 200,000 to 400,000 years ago, humans were occasionally creating material items and engaging in behaviors, such as burials, that may have reflected collective transcendent experiences. By 100,000 years ago we have clear evidence that the density and diversity of these items and behaviors had increased, and that they had become more widespread. By 40,000 years ago, evidence of directly representational art can be found across the human landscape, and human groups are clearly engaging in legisign-laden ritual. The density and connectivity of ideas, materials, and groups continued to increase over the millennia. We know that by 20,000 to 30,000 years ago there is abundant material evidence of meaning-making everywhere humans are. Human groups are generating more complex social structures and living in greater density. Then domestication begins, altering tool kits and lifeways and ushering in the first firm and interpretable evidence of what we can identify as systems of symbol and ritual: belief systems. The more interconnected humans' lives became, the more frequent and dense is the evidence of belief systems.

During this time we begin to see distinctive group and individual identities clearly represented in the variation of items and tools, different patterns in burials, large-scale architecture, and the representation of the human body in a wide range of styles and contexts. By at least 5,000 to 8,000 years ago, humans were participating in religious institutions as we understand them today.

Religious belief systems did not simply appear out of thin air. They evolved. Like so much else in our species' history, the origin of religion was a gradual emergence that cannot be pinpointed to one moment or event.

THE "EVOLUTION" OF RELIGION

Anthropologist Roy Rappaport tells us that to understand the nature of religious belief we must think of humans as "a species that lives, and can only live, in terms of meanings it must construct in a world devoid of intrinsic meaning but subject to physical law."[28] He also suggests that neither religion as a whole nor its individual elements can be reduced to a suite of simply functional or adaptive terms. According to theologian Wesley Wildman, abundant evidence demonstrates that religion is at least partly the product of evolutionary processes, but it is likely a mix of adaptive and "side-effect" features of human evolution.[29] And theologian Wentzel van Huyssteen muses that "religious behavior can never be disentangled from the broader issue of the evolution of the embodied human self."[30]

I agree. I think one can, and should, develop evolutionary explanations for many aspects of human religious belief. But most current proposals for the evolution of religion are too reductionist to be satisfactory.

Most such proposals suggest that religion and religious belief are adaptations generated via natural or cultural selection to help humans organize in large groups and facilitate cooperation. Others posit that the patterns and structures of religious belief are generated as a by-product of the normal functioning of the human cognitive system. These arguments see religion as sets of beliefs emerging from underlying psychological mechanisms that enable humans to conceive of supernatural agents and to believe they are real. The basic assumption is that the evolved human cognitive complex[31] makes us self-aware and allows us to attribute mental states to others that differ from our own. The argument is that such a system promotes "supernatural agency detection"—the creation of mental impressions that supernatural agents underlie observed or perceived phenomena such as lightning, death, accidents, or illness. Once we believe in supernatural agents as part of our experience, religious practices are thought to emerge as a logical outcome from such beliefs combined with evolutionary pressures.[32]

Most anthropologists and archaeologists see large-scale, hierarchical religions as a central part of the stratification of social systems and material complexity that develops in the last 8,000 to 10,000 years. Psychologist Ara Norenzayan argues that belief systems prioritizing "Big Gods" (moralizing and interventionist deities) emerged alongside the initial increases (about 10,000 years ago) in social complexity and coordination. As these populations became more complex, their belief systems, personified as deities, became more moralizing, interventionist, and powerful. The ritual complexes associated with these gods facilitated the large-scale cooperation and coordination that enabled the emergence of more complex societies (such as organized states). For Norenzayan, "Big God" religions are responsible for "Big Groups"—modern, hypercomplex social structures, including large-scale intragroup coordination and large-scale warfare.[33]

Unfortunately, many of the standard evolutionary proposals end up with a chicken-and-egg problem. For example, the material record of human evolution does indicate that the presence and structure of Big God religions are directly connected to increasing social complexity and material inequality. The Big

God religions are also quite likely driving forces behind the fostering of civil control, punishment, and intergroup conflict in the early large "state" societies. But it is not clear that they came before such things emerged or even early in the process.

Norenzayan thinks that strong cultural evolutionary processes have resulted in a system that links prosociality, morality, ritual, and "deep commitment" to the development of belief systems with Big Gods who are powerful, interventionist, and punishing, and who require hard-to-fake commitment. These characteristics, he argues, allowed the Big God religions to outcompete their rival belief systems to become the dominant religions for humans today.

Other behavioral scientists, such as Dominic Johnson and Jesse Bering,[34] have proposed similar explanations for the emergence of Big God religions, but they focus on the role of supernatural punishment as the key source of hypercooperation in human groups and of conflict between them. This proposal again ties warfare and centralized control to the emergence of big religions. In this reasoning, the major contemporary religions, with their moral policing and punishing Gods, are the direct product of natural selection for specific cognitive characteristics. In short, they argue that humans evolved mental capacities to create religions centered on big, punishing Gods in order to be able to coordinate larger and larger social groups. These arguments have become very popular and seem to make some sense. But they raise a few problems.

The Big God scenarios overemphasize the need for human communities to develop new methods of coordinating cooperative interactions at a large scale. Humans were already practicing intensive cooperation well before the advent of full-blown agriculture or sedentism or increased social inequality. And this infrastructure was certainly well established before the time frame in which the current Big God religions appear,[35] some 5,000 to 8,000 years ago. Why didn't such religions emerge earlier? Obviously, there is a connection between domestication, agriculture, increasing inequality, and the emergence of contemporary religions. But it is hard to make the case that Big God religions evolved mainly to facilitate large-scale civil societies, even if they did play major roles in structuring and expanding some of them, when the earliest of these societies preceded the religions by several millennia.

The Big God story is incomplete at best. It is not an explanation of religious experience but of the rise of particular kinds of belief systems and institutions. But in order for such complex and coordinated religions to emerge, religious experience must already be firmly established as part of the human landscape.

There are other evolutionary arguments, beyond the "Big God" and supernatural agency detection hypotheses. Psychologists Pascal Boyer and Brian Bergstrom describe "being religious" as the performance of rituals more or less directly connected to beliefs about nonphysical agents.[36] They argue that meaning-making activities and concomitant ritual behavior become common and central in the human experience, and that this then catalyzes the emergence of more formal religion.[37]

Good evidence shows that ritualistic behavior is closely associated with toolmaking. By 300,000 to 500,000 years ago, the ways in which members of genus *Homo* were constructing, distributing, and using stone tools suggests a behavioral complexity that is patterned like ritualistic behavior.[38] We do see inordinately symmetrical and highly aesthetically pleasing tools in the material record starting as much as 500,000 years ago. Why do so much extra work if the function of the tool does not rely on it? Practical tasks, like making stone tools, may have assumed a ritual aspect that is familiar to anyone who practices a traditional craft today: one does things "the right way" for reasons that have as much to do with connection to a tradition and community as with utility. Such feelings eventually allowed ritualized actions to extend beyond functional activities like toolmaking into other meaning-making enterprises. The timing of the appearance of meaning-making and the evidence of legisigns does offer some support for such assertions. Some researchers point to increasing evidence of specific neurological structures associated with prelanguage communication and skill transfer to suggest that ritualized behavior played (and still plays) a core role in human evolution.[39] They suggest that rituals in practical activities set the stage for the emergence of ritual generally, enabling the rise of more complex behaviors that eventually led to religion and religious perceptions and experiences.[40]

Anthropologists Candace Alcorta and Richard Sosis suggest that the differentiation of religious ritual from practical ritual can be seen in the emergence of emotionally charged symbols.[41] They argue that a key aspect of religious ritual is that it creates the possibility of a transcendent experience. It is genus *Homo*'s developmental brain plasticity and extended childhood, they suggest, that make humans highly susceptible to emotional priming, especially when we begin to be involved in the creation and participation of meaning-making systems. In this scenario, religious experience emerges from human capacities for ritual, meaning-making, and emotional engagement and is co-opted by evolutionary processes.

This argument puts transcendent experiences at the core of individual believers' experience of what it is to be "religious." The cognitive, physiological, and perceptual realities of religious individuals vary, and that variation may contain a complexity that is missed when we focus on what being religious *does* rather than what it *is* for believers.

Yet the majority of evolutionary explanations focus on what religion "does." This is where the fault lies in these approaches; they largely ignore what the religious experience *is* for believers. I don't think we can disregard the experience of *the religious* in favor of structural and functional evolutionary explanations for religions or religiosity. The concept that being religious is primarily a by-product of the complex human cognitive system, and/or that being religious is an adaptation to enhance human cooperation and advance social complexity, is too shallow a way to think about religion, evolution, or the human experience.

IMAGINATION → BELIEF →
RELIGIOUSNESS → RELIGIONS

Anthropologist Barbara King suggests that emotional and social bonding in primates provides the infrastructure that allows humans to be religious.[42] As human ancestors evolved, the neural and social complexity needed to think beyond their immediate experience, combined with their strong sense of belonging, expanded into an infrastructure for religious belief. Neuroscientist Patrick McNamara argues that the capacity to be religious is a central facet of the human development of "self."[43] Theologian Wentzel van Huyssteen suggests that the human religious imagination is part of the processes that facilitated human evolutionary success over the past few hundred thousand years.[44]

Van Huyssteen recently summarized my own take on the evolutionary basis for religiousness better than I do: " Agustín Fuentes and I, in spite of our radically different disciplines, approaches and methodologies, completely agree that a necessary prelude to having religion is indeed the emergence of a human imagination and the embodiment of a quest for meaning as part and parcel of the distinctive human niche."[45]

I propose the following as a complement to the eloquent analyses of many scholars on this topic and offer it as my expansion on the evolutionary framework for religious belief.

By 1.5 million years ago, the human niche included the creation and use of tools, developing techniques with a level of complexity that no other creature on this planet has ever matched. These tool-creating, -making, and -using processes and their successful entanglement to early *Homo* lifeways created a feedback dynamic. Changes in motor, cognitive, and behavioral skills grew intertwined with creating, using, and sharing the practical, social, and mental aspects inherent in these tools. Around the same time, *Homo* shifted toward novel body and brain growth patterns that included a greatly elongated childhood. This was both necessitated and facilitated by shifting patterns of caretaking, increased quality of diets, and an expansion of social coordination and communication.

A million years later, tool creation and use, elongated development, and intensive social cooperation had become embedded in the genomes, bodies, and minds of genus *Homo*. They continued to expand and become more complex. By 300,000 to 400,000 years ago, their lifeways included fire, increased innovation, the creation and use of more intricate tools, large-scale hunting and more diverse resource exploitation, expansions of ranges and ecologies, and the initial glimmerings of meaning-making. This timing also maps to the solidification of the auditory and vocal anatomy necessary for language, and to the brain reaching its contemporary size. Emerging from the confluence of all these processes is a robust and essential human imagination—a capacity to engage daily life with cognitive processes that are not limited to the here and now, to the material at hand, or to one person's specific experiences.

After more than 1.5 million years of evolution, the human capacity for imagination that is "stupendously vast, stretching across the real and the unreal, the possible and the impossible"[46] generates a consistent ability to experience the transcendent. Transcendent experiences become a central component of the human niche.

Over the last 300,000 years or so, the evidence for the growth of the transcendent is increasingly abundant. Legisigns emerge, and their replicas appear more and more frequently. Tool kits complexify faster and in more directions. Human groups move more widely and commingle more extensively, creating new ecologies, societies, and possibilities, continuously adding fodder for the imagination and expanding the landscapes of belief.

Human groups begin to use aspects of meaning-making as central facets in their lives. The emergence of language and other symbols, and the myriad ways groups create new and deeply held senses of identity and community, got fully under way. During this period, the transcendent and the imaginary become as central in the lives of humans, and as evolutionarily relevant, as the physical and the social. Drawing on cognitive and social resources, on histories and experiences, and immersing themselves in transcendent engagement enabled a new landscape of beliefs. Humans began to "see" and feel and know that their world contains more than the material at hand. They became religious.

By "religious," I mean that they used their capacity for belief in the transcendent to establish powerful, persuasive, and long-lasting moods and motivations. These need not have been tied to specific formal doctrines, practices, texts, or institutions. From the archaeological record we can clearly see that beginning by at least 200,000 to 300,000 years ago, imagination becomes prominent in humans' lives in a way that we often associate with religious experience. The centrality of religious capacity as a feature of humanity solidifies.

Between about 15,000 and 4,000 years ago, we see another set of radical transitions: the advent of plant and animal domestication, storage capacities, and concepts of property. Inequality emerges and increases within and between groups, with increased sedentism, and the growth of towns and eventually cities—all of which leads to the formation of large-scale, multi-community polities with multilevel political and economic structures. These changes bring a restructuring of human lives and societies with the emergence of a range of institutions and other larger-group social structures. The restructured lives and societies now include, shape, and are shaped, intensively, by human beliefs and the capacity for religiousness. We begin to see, especially toward the end of this period (some 4,000 to 8,000 years ago), material evidence for the emergence of formal coalitions of religious beliefs and practices, as well as the material symbols and structured institutions that unite them under a specific theological doctrine and ritual. In other words, we see religions.

That is how I think the capacity for religion evolved and religions emerged: as facets, patterns, and key processes of the human niche.

This explanation does not tell us why any given religion has the particular practices and beliefs it does. That is because it can't. Unlike many of my evolutionary explanation–oriented colleagues, I'm fully comfortable leaving open the possibility that some form of transcendent revelation plays a role in a religion's particular beliefs. As a scientist I can rule out theological assertions that contradict material facts. For example, the earth *is* really, really old, and there was neither a physical Garden of Eden nor two original progenitors for our species. But I cannot assess truth claims about sensations and beliefs based on faith and transcendent experiences. The insights given by deep faith and devotion are real for billions of humans today. Billions. Cultural constructs, regardless of their origin, are real for those who hold them. Belief in a set of faith practices, rituals, and theology is a commitment that humans have evolved to make.

Meaning-making, the transcendent, and openness to revelation and discovery are core parts of the human niche and central to our evolutionary success. They are why we believe in the ways we do. These capacities have been and continue to be a good thing for humanity. But they also introduce horrible possibilities.

Economies

MANY PEOPLE believe contemporary economies emerged as a necessary reality of the world.[1] This is at least partially wrong.

Nearly all humans participate in some form of an economy—an organized system of activity involving the production, consumption, exchange, and distribution of goods and services.[2] We often work to earn compensation and then use that compensation to purchase items and goods that are not the direct products of our labor, such as food, housing, or transportation. If we don't have money or credit, we can try to trade some form of labor for a good, such as fixing someone's roof in exchange for food. If we have some goods but need others we can also barter, exchanging our goods for other goods.

Today, the distribution of wealth in economies is rarely equitable, especially in market economies, the dominant form of contemporary economy. In the first quarter of the twenty-first century, a tiny percentage of the more than 7.5 billion humans on the planet control the vast majority of wealth in the global economy and in almost all local and regional economies.[3] If contemporary economic systems are so inequitable, why are they the dominant pattern for human societies today?

In part, because of belief.[4]

A wide swath of humanity believes that economies, especially market economies, are natural phenomena, like language capacity, complex neurobiology, and the use of symbols—universal and inevitable outcomes of being human. Across many societies[5] there is a shared belief that every benefit has a cost, that there is no "free lunch," and that making a profit, making a "good trade" and "coming out ahead" are not just what everyone seeks but are reflections of the natural pattern of life. In short, many people believe the world is a system in which competition for limited resources is very close to a natural law. The astute assessing of the costs and benefits, and an eye for greater overall benefit from economic exchanges, makes one a winner in the game of life. This belief in the "naturalness" of economic processes includes the assumption that humans have a natural drive to negotiate, to exchange, and to barter. It presumes that humans are economic actors and that market economic systems are therefore the natural outcome of our evolutionary trajectory.

It turns out, however, that most humans do not behave like rational economic actors. We willingly accept losses as often as gains in exchanges. The reason is that for the majority of humans, exchanges are not about profit but about making and keeping social connections. The relationships at the center of human social lives are not driven by an urge to obtain equal or better benefit in interactions. Think about the relationships between mothers, fathers, and children; between loved ones and best friends; or even about your "investments" and "returns" in your dog or cat. What you give, get, and share in these interactions cannot be quantified or monetized. There is much more to human social lives than can be modeled in economic cost/benefit calculations. And we are not alone. Ample research demonstrates that other animals, plants, and even microorganisms do not always, or even mostly, behave as if the world were a place of dire competition where success means making an evolutionary profit.[6]

But if the world is not one big economic system, why do so many of us believe that economic processes are "laws" of nature equal to gravity, evolution, and thermodynamics? Why do so many humans believe that economies reflect a specific and inevitable outcome of human nature?

HUMANS CREATED ECONOMIES

The infrastructure to believe in economic systems is a product of our evolved capacity for imagination, creativity, and language and our devotion to creating, modifying, and sharing information and material goods. These capacities and tendencies have become central features of the human experience and the human niche. They allow us to both develop economic perspectives—to see and believe that the material goods and knowledge we produce are both valuable and transferable—and to integrate that perspective into our perceptual landscapes and worldviews. Key among the economic beliefs we create are assessments of "value" and convictions about how that value should be recorded, exchanged, and managed.

We believed economies into being. No other species has the capacity to do this. But why we believe what we believe about economies today, why specific types of economies dominate our beliefs and lives . . . that is another story.

Our tendency to see economic patterns and relationships as necessary and central elements of the world comes from specific patterns in human evolution and history over the last ten to fifteen millennia. In addition, many humans believe certain things about specific economic systems because of very recent events, processes, and patterns in the human experience.

The most recent segments of our evolutionary history enabled the development and dominance of a particular type of economic perspective. Storage, surplus, concepts of property, role specialization, expansion in the size and density of human settlements, domestication, and even warfare reshaped the human material and social world in such a way that economies[7] emerged as central features. Within the last few thousand years, changes to human political, demographic, philosophical, and religious systems enabled a certain type of economy, market economies, to arise. Market economies are money[8]-based systems where investment in, production, and distribution of goods are guided by price valuations largely set by the availability of supply and the rate of demand.[9] Over the last few centuries, historical, political, and philosophical ideologies came together to develop these economies' current variants (primarily capitalist[10]) and their alternatives (command and mixed economies).[11]

While economies do not reflect "laws" of the natural world, they have emerged as a deeply rooted and nearly ubiquitous feature of human societies. To better think through why many *believe* modern economies reflect the natural order of things, it is critical to elaborate on what an economy actually is versus what we believe it to be.

What Is "an Economy"?

In the traditional academic sense, "economics" is the discipline concerned with understanding the production, consumption, and transfer of resources. But it is much more than that. Economics is a philosophy[12] as much as a social science. The American Economic Association tells us that it is "the study of scarcity, the study of how people use resources and respond to incentives, or the study of decision-making. It often involves topics like wealth and finance, but it's not all about money."[13]

For economists, "economics" is an approach to modeling and understanding how the world works in the same way biologists study biological systems and physicists study the patterns and processes of matter and energy across space and time. Traditionally, economists began by borrowing from philosophers and assuming that the human capacity for rational decision making enables us to attempt to maximize utility as a consumer and profit as a producer—that humans make rational economic choices when it is in their self-interest.

One could rightly argue that because all humans use and exchange resources, we are all entangled in economic systems— and that these systems do have predictable processes and patterns. But few biologists and physicists think economic processes and patterns are directly comparable to the physiological and ecological processes that affect biology, or to the physical and thermodynamic processes that affect the interactions of matter and energy. Nor do most social scientists see the study of economic models as akin to the study of matter or living systems. To be fair, many economists agree that in studying economies they are assessing human social systems that have emerged over our social and political histories and that economic models don't necessarily reflect an underlying biological or evolutionary history. But others disagree. They see their economic models as describing a deep evolutionary and psychological reality, and they believe that economic processes and patterns are a mirror of human nature.[14]

What are the core assumptions of economic models? And can they tell us anything about why we believe in economies?

A good place to start is with Scottish philosopher Adam Smith, often called the father of free-market economics. While Smith wrote broadly on a range of topics,[15] he is best known for the economic theory he developed in his 1776 book *An Inquiry into the Nature and Causes of the Wealth of Nations*.[16] In essence, Smith proposed that rational self-interest, combined with a "free" economy (where the government or other controlling interests did not set prices or control production) can lead to social and economic well-being. He believed that human development could reach its best outcomes in environments of open and free economic competition that operated in harmony with what he considered natural laws.

A person earning money by his own labor, Smith wrote,[17] benefits himself but also benefits society, because to earn income on one's own labor, one must produce something others value.[18] And although one's intent in creating and delivering products of value may be purely to advance one's own benefit, one is actually contributing to the betterment of all. Such actions foster a system in which the best products, the best labor, and thus the best patterns of exchange all rise to the top, without the need for intentional coordination toward this goal by those involved. What Smith famously called an "invisible hand," the cumulative action of individual self-interest, would guide the market toward this state.

Smith believed that social harmony could emerge naturally from a free and open economic system. The same assumption undergirds contemporary free-market ideologies and forms the basis of much of the global economy.[19] It also provides the core infrastructure of what many of us are taught is how economies "naturally" work.

In the twentieth century, economic theory was further shaped by the work of John Maynard Keynes.[20] Like all economists, Keynes was concerned with the problem of demand and supply, but he viewed it at a national rather than a local or individual level. He proposed that when demand falls short of productive capacity, unemployment and economic depression result. But when demand exceeds productive capacity, it brings inflation.[21] In his models, Keynes tried to demonstrate that purely open or free markets, if left entirely unmanaged, are insufficient to maintain a balanced ratio of demand and supply, and that there is no automatic tendency to produce at a level that results in full employment of all available human capital. Therefore, he argued, governments and other significant financial actors have to participate in the structuring of markets to achieve the desired results. His approach modified Smith's basic notion by adding that some structural controls have to be in place. Today, many economists are critical of Keynes; however, it is universally understood among them that there are frictions in markets, and as a result real-world markets, left to their own doing, do not necessarily generate the most efficient outcomes.[22]

The third major economic figure of relevance deviates significantly from Smith and Keynes: Karl Marx.[23] A political philosopher, Marx asserted that it is systems of labor, production, control of compensation, and the taking of profits that dictate the majority of economic outcomes. He asserted that within the capitalist system (the dominant market economy in the nineteenth century as well as today[24]), labor was not equitably connected to income or product. Rather, it was a commodity that in a completely unregulated market could gain only subsistence wages. Capitalists, those controlling the systems of production, could force laborers to exert more labor than was necessary to earn basic subsistence.

These capitalists could take the extra product created by the workers and profit from it, without sharing that profit with the laborers. Marx believed that markets and capitalist practices pushed wages down and that the value assigned to goods and services did not accurately account for the true cost of labor. An economic system based on markets and private profit, he argued, is inherently unstable because of the abuse of the laborers and the unfair distribution of compensation and profits. Capitalism in his view was driven by a deeply divisive class struggle in which the ruling minority appropriates the surplus labor of the working majority as profit.[25] Marx's ideas were a basis for the one major economic and political system that opposed the capitalist system: communism. By the twenty-first century, however, that system had failed.[26]

Most importantly, Marx reminded us that economies are not "nature" but human creations. He wrote, "Economists express the relations of . . . production, the division of labor, credit, money, etc. as fixed, immutable, eternal categories. . . . Economists explain how production takes place in the above mentioned relations, but what they do not explain is how these relations themselves are produced, that is the historical movement that gave them birth. . . . These categories are as little eternal as the relations they express. They are historical and transitory products."[27]

This is a critical point: economic systems are indeed real, and they can be modeled, debated, theorized, and described. But they are not the results of natural laws. They are not naturally emerging features of the world. Economic systems and ideologies are not some inevitable product of unchangeable human nature. They are human-made, certainly creative and imaginative and very real, but the products of human society. They exist because we created them, and they are maintained because we believe in them. As philosopher Ian Hacking puts it, economic belief has "looping effects": believing in money actually makes money happen, with effects on human behavior that make money a part of people's shared worlds with real effects. Anthropologists James Carrier and Daniel Miller highlight a critical problem that this belief creates in studying economies: in practice, economic models are not actually measured against the world they seek to describe, but instead our world is measured against them.[28] Such an approach can make it look like the world does function, naturally, as an economic system.

Luckily, today many economists and other social scientists don't believe in a purely economic nature of reality. Adam Smith's rational actor—laboring in an open, competitive market to produce the best possible product at the best possible price and to be equitably compensated for it—is an admirable philosophy but not a reality. That individual economic choices are made independently of local and global processes and structures has been substantially challenged by Keynes, Marx, and many others. Economist Carlos Alós-Ferrer tells us that "in spite of popular beliefs, economics has already moved past the neoclassical dream of all-knowing, optimizing, self-interested agents. Several decades have passed since the first microeconomic models incorporating bounded rationality found their way into the mainstream. . . . Behavioral economics challenged the self-interest assumption and started opening the black box of human motivation"[29]—a "black box" that is structured by belief.

So, if many economists today agree that economies are products of human creativity and not matters of immutable natural law, why are simplistic beliefs about the economic nature of humanity so pervasive? Why do so many people believe humans are "utility-maximizing rational actors," to use the economic jargon, who seek to maximize utility as consumers and profit as producers, when economists themselves mostly no longer do?

What Humans Believe about Economies Is Influenced by What We Think "Nature" Is Like

One can define an economy as an organized system of activity involved in the production, consumption, exchange, and distribution of goods and services. One might argue (and many do) that we can see the rudiments of such systems in the natural world. Bees, birds, salmon, gophers, giraffes, dung beetles, and all other living animals all seek out, identify, and consume items that offer benefits (energy); they then turn those benefits (via physiological labor) into a product (offspring). The items that offer them energy are generally assumed to be (a) not ubiquitous, (b) not free of effort (cost) to acquire, and (c) sought after by other competing organisms. Isn't this roughly the same as a human economy?

Behavioral ecology[30] is the area of study that focuses on what we might call the "economics" of the natural world. Primarily, behavioral ecology looks at how competition and cooperation between individuals and species affect "fitness," meaning evolutionary success.[31] Behavioral ecologists generally apply cost/benefit analyses[32] to the relationships among organisms, their ecology, and their behavior in describing and assessing their behavioral and ecological strategies (how they "make a living"), and how and why these strategies evolved. The standard behavioral ecology model borrows from aspects of human economic models, particularly market systems, in that the evolutionary "value" of a behavior is assessed based on its potential profit or cost relative to the goal of successful reproduction[33].

This approach recognizes that organisms seek to acquire or develop patterns, partners, traits, and behaviors that will help maximize their capacity to produce offspring. It's assumed that there is a kind of arms race in acquiring these traits, so that any trait that confers an advantage and spreads through a population will eventually be matched or exceeded by a different trait developed by others competing for the same resources and reproductive opportunities. In fact, one prominent current approach is the application of "optimality models"[34] to the processes by which organisms develop their collections of adaptations. These models do not assume all organisms are optimally adapted (or rational), but instead set up a suite of assumptions for how organisms should behave and adapt if they are driven to achieve the best evolutionary outcomes. These models employ nearly the same assumptions that underlie contemporary free-market economies. Then, when these outcomes are not reached (as they almost never are in the real world), the researchers can look into the system and try to find the constraints that inhibited the development of optimal outcomes. The assumption here is that evolutionary processes "strive" for optimality, even if they rarely achieve it.

The comparison of natural systems to market economies has gone even further. Biologists Ronald Noë and Peter Hammerstein[35] have proposed what they call "biological markets." In this approach, individual organisms exchange commodities (such as behavior or items of value) to their mutual benefit, and the exchange value (the natural equivalent of money) of these commodities is a source of conflict. In these exchanges, behavioral mechanisms such as partner searching, partner choice, and contest among competitors determine the composition of trading pairs or groups. Noë and Hammerstein explicitly use the term "biological markets" because they see a direct analogy with human markets.

But is this an accurate depiction of how organisms actually live? Research from across multiple disciplines shows that models rooted in economic assumptions have only limited ability to represent complexity, history, and evolutionary processes in many organisms.[36] The cost/benefit analyses can help us understand some aspects of decision making and behavior, but they are a poor representation of the processes of life.[37] Optimality models make a suite of assumptions, based on the cost/benefit approach of behavioral ecology, and in most cases produce mathematically plausible outcomes while telling us little about the actual evolution of systems or organisms.[38] In short, while mapping economic assumptions onto the natural world helps humans (especially academic researchers) develop models that we find meaningful and understandable, it does not show us that the natural world functions as an economic system.

So why do many scholars find economic models of the natural world so attractive?

Biologist Hannah Kokko[39] helps us think through this by looking at what kinds of questions researchers ask about other organisms. For example, when behavioral ecologists ask, "Why do male birds often continue to provision young at the nest even though some of the offspring are probably fathered by someone else?" they often go straight to an economic model for an explanation. "All that is required," Kokko tells us,

> is to imagine a dilemma with trade-offs, perhaps with multiple players with divergent interests, and hope for honest self-inspection: "what would you do?" Although our mind prefers shortcuts and heuristics over painstak-

ing calculations of all relevant probabilities, our ability to intuit scenarios involving personal gain, weighing the relevant pros and cons, is decidedly better than, say, our ability to truly grasp quantum physics or cosmological timescales.

It is much easier for many researchers to understand a model of the world using economic terms and processes, such as calculating the relative cost/benefit for male birds to feed others' young as long as their own young do not suffer and there is a net benefit to the male bird. But it does not explain why this behavior appears in this bird. It's just that it's easier for those of us steeped in these scholarly traditions, and their economic assumptions, to comprehend such scenarios when we describe them in those terms.

Why? Because so many of us are trained in a system that instills deep belief in economies as natural. The concept of the personal benefit outcome–oriented, decision-making human is deeply ingrained in the communal psyche.[40] So it's easier to think with that model. As I noted in chapter 6, cultural constructs are real for those who hold them. Storage, surplus, concepts of property, role specialization, expansion in size and density of human settlements, domestication, and the recent emergence of contemporary economic systems have shaped ways of thinking about the world so that perception of these economic systems as a natural reality is increasingly easy, and common, for much of humanity.

But even as many of us believe there are laws and natural patterns governing the production, consumption, and transfer of wealth, in much the same way that the law of gravitation governs falling objects, many humans don't actually behave as if the world functions economically. And that matters.

HUMANS ARE NOT ALWAYS GOOD
ECONOMISTS

What do people want? Are we primarily seeking outcomes that benefit us economically? That is certainly what popular economic ideologies imply and what many people believe, but is it true? Starting in the 1990s, a large group of psychologists, economists, biologists, and anthropologists gathered[41] and asked these questions about fifteen small-scale societies[42] in twelve countries on four continents. The group sought to use economic experiments and ethnographic investigations to directly assess people's patterns of choice and decision making in economic contexts. They wanted to study economic choice behaviors in a range of societies that are not substantially tied to contemporary urban infrastructures or immersed in market economies to see if being outside those contexts produces different outcomes in classic economic testing scenarios.

They did not find a single society in which people consistently behaved according to the expectations of basic economic theory. They found a huge amount of variation in how the different groups responded to economic options. The degree to which groups were integrated into and reliant on market economies, combined with the importance that their local societies placed on cooperation, explained most of the different responses to economic experiments put to the various groups. That is, those most connected to the obligatory use of money and the purchasing of basic items in stores, rather than building or growing them themselves, showed the closest outcomes to peoples from Western market economies (such as the United States or England). Neither individual patterns of choice and preference nor differences in demographic patterns among groups explained the variation in responses between the groups. Most interestingly, a majority of the behaviors people displayed in response to the economic experiments mirrored the ways people interacted in their daily life in their own society. In short, the results tell us that (a) being in a market economy shapes people to behave in ways that mirror how the economy is structured, (b) there is no one "naturally human" way to participate in economic exchanges, and most importantly (c) culture matters.

Since the original study was done, multiple research groups have validated these findings many times.[43] Our fixation with economic models of how humans get along is the product of recent cultural history, not something deeply rooted in our evolutionary trajectory. But that doesn't make it any less real for those humans who believe it and experience it.

The notions that undergird theories about economies and their expectations did not emerge from thin air. They have their roots in material and social processes that humans created, experienced, changed, and continue to develop. Two key transitions in our evolutionary histories, central to this pattern's emergence, are keys to understanding why we are prone to believe in economies. They are, first, the increasing reliance on the creation, maintenance, and exchange of extrasomatic (not of the body) material items as central to human lives and ecologies, and second, the move from materially egalitarian systems to systems of ownership and property, which catalyzed the emergence of the first *actual* economic systems.

WERE HUMANS EVER REALLY EGALITARIAN?

While there are (or were until recently) small-scale societies with little economic role differentiation and largely communal sharing of goods that are often labeled "egalitarian," one must be careful using this term as it can convey substantial misperceptions. "Egalitarian," applied to human social structure, is not a statement about equal rights but a practical description of social and economic practices. It does not imply a utopian past free of discord or conflict. Earlier humans did not spend their lives running through fields of daisies holding hands. They disagreed, sometimes violently, over access to resources, feelings of inequity, and questions of leadership, even as their social world remained centered on sharing of goods and equality of power.[44]

Anthropologist Polly Wiessner[45] does a stellar job of explaining key elements of egalitarian human societies via her studies of the Enga peoples of New Guinea. She shows that egalitarianism is not the product of simplicity and is not humanity's default social structure. Rather, she argues, "to live as a member of an egalitarian society is an active process" amid varied and complex social structures. Egalitarian processes are not simply pared-down versions of hierarchical power structures, nor are they the product of a lack of material ownership. They are rules and expectations maintained by the relationships, histories, and beliefs of the society. The egalitarian societies we've been able to study all maintain socially defined distinctions of age, gender, ability, and kinship, even while they hold egalitarian expectations.[46] Rights and access to social position are negotiated, developed, and maintained by social relationships. Material and social egalitarianism is an actively created and negotiated cultural system.

It's specifically because of the active structures and dynamics of egalitarian societies that humans were able to develop and expand the possibilities of inequality—the first step toward modern economic systems. Wiessner tells us that "the strength and configuration of these coalitions [the active nature of egalitarian relations], together with ideologies of what constitutes a transgression of the norms of equality, produce a wide range of variation in so-called egalitarian societies" and that can create "pathways to inequality."[47] Anthropologist Christopher Boehm adds to this by demonstrating that in many egalitarian societies, shared or communal property is socially defined and actively reinforced, not an automatic process resulting from a lack of goods or individual inability to maintain or defend them.[48] Humans make and share materials more than any other species. This is a starting point for the emergence of distinctively human economic processes: the intentional and socially defined sharing of common goods, and the social mechanisms for doing so.

A diverse body of data supports the assertion that most human populations practiced a general pattern of economic egalitarianism[49] until about 12,000 years ago.[50] Why have we changed since then?

WHY HUMANS HAVE THE INFRASTRUCTURE TO BELIEVE IN ECONOMIES

As described in chapter 4, starting about 12,000 years ago, many human populations became more and more committed to sedentism and domestication, developing tendencies for monumental architecture and larger and larger communities.[51] Social structures were developing and expanding in a way that created new opportunities for social difference and inequality.

In villages, then towns and cities, there was a move from domestic storage to public storage. Public storage needs new processes of collection, management, and distribution, leading to the emergence of an elite social cohort with different responsibilities from the rest of the population and new patterns of exchange of the stored goods. Increasing complexity in systems of storage, surplus, and management; the creation of elites; and the collection and redistribution of goods is observed in the remains of many human populations around the planet over the last 4,000 to 8,000 years as populations have increased in size and complexity. We also begin to see clear evidence for contemporary patterns of gender differences. And at the end of this period, we get some of the first evidence of written records, the earliest of which record the exchange and monitoring of goods.

With the rise of towns and cities comes a diversification in the material needs of their inhabitants and thus the emergence of role specialization and of craftspeople. This specialization increased people's reliance on material items and also led to diversification in the social structures for the production and management of these goods. The need arose for different training for different skill sets, as well as differential access to specific goods, which set the stage for extensive, hierarchical, and complex economic systems.

These new, more diverse material economies and differentiated population structures enabled new types of political organization. As noted in chapter 4, societal elites eventually became more than merely managers of storage and surplus. They began to oversee the movement of people and goods and the distribution of labor across the society.

It is in the past 5,000 years or so that clear evidence for contemporary economic patterns shows up, including patronage, the exchange of goods for labor, the emergence of market-based economies, and the development of currency systems. This changed what, and how, humans living in these societies believed about value, exchange, equity, and each other.

A recent proposal holds that this pattern of developing inequality and its related economics have begun to shape the human niche—a major force in human evolutionary landscapes—and that this is one of the reasons so few materially egalitarian societies are left. Anthropologist Siobhán Mattison and colleagues argue that while inequality likely emerged through a variety of local processes and pathways, "Its evolution is fundamentally dependent on the economic defensibility and transmissibility of wealth."[52] Economic defensibility, the ability to successfully protect your goods and resources from others, they argue, emerged because of increased resource density and larger populations. At the same time, agriculture and other domestication were making resources both more predictable and more abundant, enabling the transmission of wealth across generations. The possibility of secure inheritance of resources and goods greatly raised the importance of material wealth versus other kinds of wealth. It also allowed much greater wealth accumulation. Property rights became institutionalized. Persistent inequality in material wealth brings patronage, extremely hierarchical political leadership, managerial elites, and permanent class divisions—the basis of contemporary economies and economic thought.

TRADE, MONEY, AND WEALTH

Trade is old. There is solid evidence that by 400,000 to 300,000 years ago, material items for stone tools were being transported as much as 50 to 100 kilometers, possibly as items of trade between groups.[53] By 50,000 to 100,000 years ago there is clear evidence that items are transported even larger distances, and trading was the likely cause. Trade is a significant factor in the evolution of contemporary human social systems.[54] Nearly all human groups have relationships of exchange with other groups. But "exchange" and "trade" are not necessarily related to our contemporary economic beliefs, and in fact their origins are quite different.

French sociologist Marcel Mauss argued that the exchange of objects builds relationships between human groups.[55] Rather than envision such exchanges as economic, with values being assessed and monitored, he suggested that the reciprocal giving of gifts was a social act, not an economic one. The act of gift giving created relationships between groups and enabled social trust, mutual entanglement, and patterns of bonding that facilitated intergroup affiliation. We know that the deep cooperative and social nature of the human niche establishes this social pattern as a baseline human process, where the exchange of gifts is a regular part of the human mode of being in the world.

Anthropologist David Graeber[56] reinforces this perspective by offering ethnographic and historical evidence that many (or even most) human exchanges are not seen as an economic relationship but as a mode of sociality—a way to connect, without specifically accounting for the goods' value or seeing the interaction as a transaction with costs and benefits. Gift giving, sharing, and the exchange of materials between individuals and groups appears to be a central feature of human social life. In most cases of such exchange, strict reciprocity[57] is not expected. In economic systems, by contrast, accounting is critical. Once exchanges become subject to strict valuation, and an accounting of that value is introduced into the transactions, an economy is started and obligatory reciprocity emerges.

How did we go from exchanges as social relationships to exchanges as economic ones?

The common story, told by many economists and historians, is that once humans began to develop and trade material items, a system of barter developed. Imagine group A has items x and y, and group B has item z, but when used together, items x, y, and z are more beneficial than when used separately. Therefore group A wants item z from Group B, and B wants x and y from A. Under the standard story, trade and barter started with exchanges of items (supply) that were needed or desired by those who did not have them (demand). Once this accounted exchange begins, those involved start assigning value to the items so that exchanges (barter) become a negotiation over the relative value of the items being exchanged. One can see that it's only a few steps from a barter system to a money economy. Once a system of value is agreed on, then tokens representing specific values (money) can be used in the exchanges rather than directly exchanging the items themselves. It is easier to carry around a bag of coins than a herd of cattle.

This story is convincing, but it's not true. Barter systems did not precede monetary systems; they coevolved. In most cases in the world where barter is common, it exists as a by-product of the cash economy, not as its underlying cause.[58] The "money" in such economies is a very complex topic[59] and a very recent human invention. In the most general sense, money is some form of token or record that people accept as payment for goods or services, or repayment of debts, because they have faith that others will accept it in turn. To function as money, a thing must serve as a medium of exchange, a measure of value, and a means of storing value.

But even to create the concept of money, one must have already developed some valuation system for exchanges and some way of accounting for that valuation. The archaeological evidence for anything like a money or barter economy is nonexistent until the last 5,000 to 7,000 years at the earliest. In other words, the entire system of cash-based market economics that most people take as "what humans do" was not present at all until very recently in our evolutionary history.

To develop a contemporary economy, one has to start with material inequality, some degree of differential access to items within and across groups. The earliest evidence for such inequality is about 30,000 years ago[60]—but only in the context of grave goods, and there is no discernible pattern in who gets the greater grave goods. This evidence of early inequality is not consistent with contemporary economic models (and is more in line with Mauss's social model of exchange). To develop an economy, especially a market economy, there must be some consistent and predictable way of differentially acquiring and maintaining wealth.

The term "wealth" is usually associated with material prosperity. In economic terms, it means a disproportionately large possession of goods or the power to obtain, produce, and control them. Being wealthy generally means having a lot of goods or money relative to others—a state not possible without a degree of material and social inequality, and therefore a way of being that was *not* available to humanity until very recently.

But in thinking about the evolution of inequality, Siobhán Mattison and colleagues expanded the definitions and notions of wealth to enable a better evolutionary examination of its origins. They define "wealth" as "any of a broad array of factors that are transmitted to offspring and can affect an individual's health, well-being and reproductive success." They divide wealth into categories, including material wealth (goods, land, livestock), social or relational wealth (strong social support networks, ability to recruit many social partners in support of one's ends), and embodied wealth (skills, knowledge, and health-based differences). It is the retention and transfer of wealth that are important for the development, augmentation, and maintenance of inequality. While all forms of wealth described are heritable to some degree, material wealth "is especially conducive to supporting inequality" and thus central to the development of contemporary economic systems and beliefs.[61]

By 8,000 to 12,000 years ago (maybe a little earlier), we start to see storage of goods and surpluses of material items at larger scales in villages, towns, and eventually cities. By 5,000 years ago we see clear evidence of accounting systems, substantial exchanges of material goods, and increasingly large-scale conflict between groups. Differential material wealth—the disproportionate possession of goods, or the power to obtain, produce, and control them—becomes nearly ubiquitous across human populations over the past four to five millennia. This institutionalization of a particular kind of economic inequality as "typical" in human societies is very recent, but one can easily understand how it has shaped the way most people view the world. It has created a pervasive ecology and context for human experience. We are who we grow up with, and as wealth inequality becomes a normal part of everyday life for most of humanity, people began to see, experience, *and believe it* as the natural state of affairs.

WHY WE BELIEVE WHAT WE DO ABOUT ECONOMIES

Genus *Homo* began our success story by creatively collaborating with one another and by seeing that the material world can be modified and altered to suit our needs. Creating and sharing goods is a deep part of the human experience. The dynamic patterns in humans' materially egalitarian social systems offered the possibilities for expansion and the development of more hierarchical relations. The radical expansion in our abilities to make and use tools and other items; the more recent development of domestication, sedentism, and storage; and the expansion of human groups into towns, cities, and nations brought societies into novel economic ways of interacting. These factors eventually led to contemporary economic systems: markets, money, and barter. Humans changed their own environments in ways that made egalitarian societies less and less common, and the contemporary reality of market economies became the context in which most humans develop. Growing up in a market economy dramatically influences the way we perceive the world and each other. It shapes our beliefs.

This complex set of patterns, processes, and histories tells us why we believe in economies. It is why so many humans see our contemporary economic system as natural—the way the world works. But an astute reader will notice that there's a difference between basic economic processes—the creation, movement, and management of goods in a system that has some inequality—and contemporary market economies. The degree of inequality generated by our current economic system is extreme[62] and possibly unsustainable. Our current reality is as much a product of evolutionary history as a product of recent political and economic history.

The economic systems we live in today are human creations based on assumptions and ideologies (beliefs) that have developed over the past five to ten centuries. As Marx reminded us, "These categories are as little eternal as the relations they express. They are historical and transitory products."[63] We made them, and we can change them. The human economic world continues to evolve, but that evolution is directly connected to the processes, events, and actions of the present. What will economic landscapes look like in the future? Will there be change to beliefs about the "natural" economic functioning of the world? Will the desire, or need, for reduced inequality change beliefs about the production, consumption, exchange, and distribution of goods and services? Or will the inertia of contemporary economies and their associated outcomes continue well into the future?

CHAPTER 10

Love

F OR MANY HUMANS there is nothing more powerful than
the deep emotional and psychological state we call "love."
We believe in love, and it matters.

But "love" means different things to different people. In the
Western scholarly tradition, we tend to draw our idea of love from
ancient Greek and medieval Christian concepts. Many contemporary philosophers, psychologists, and neuroscientists identify love as a desire, an urge, or a neurochemical process. Other
traditions, from diverse societies, present a range of ideologies
connecting to love attachment, caring, sexuality, and emotional
investment.[1]

While a common definition remains elusive, there is no doubt
that the suite of sensations and experiences that are collapsed
into the term exist as a component of human bodies and lives.
Despite not agreeing on exactly what love is or how to describe it,
most humans certainly believe in it. The evidence for love's centrality in human landscapes of belief is in the trails and marks that
love leaves in our psychology, neurobiology, physiology, behavior,
writings, songs, histories and myths.[2] But this "love" refers not
only to the contemporary notion of romantic love, which itself

is a concept whose historical presence across human societies is much debated.[3] Rather, it refers to the amazing human capacity to experience a particularly deep attachment and devotion to specific others.[4]

Human love can be romantic, familial, and chummy, and can even bridge species. In all cases, it involves a deep psychological, even physiological devotion. Most humans feel they've experienced some aspect of being in love and think of it as a transformative experience. But in reality, there is no single description, experience, or even a particular physiological or psychological process that we can pin down as the source or embodiment of "love." Love is not an entity unto itself but a suite of connections between human social and physiological processes, intertwined with belief. Love draws on our cognitive and emotional capacities, our experiences and perceptions, and it has its roots embedded in our distinctive evolutionary histories. The capacity to experience love deploys belief in one of humanity's most powerful contexts: interpersonal relations.

WHAT IS LOVE?

Because most evolutionary research into love takes place in Western scholarship traditions, I root my discussion, at least at first, in that context.

The ancient Greek philosophers provided three relatively distinct notions that today we might group under the rubric of "love": *eros*, *agape*, and *philia*.[5] "Eros" originally meant a passionate desire for a person or object. Typically, today, it is thought of as centered in the relations of sex and sexuality and thus is often assumed to be biological. Eros, then, is an aspect of the capacity for love that highlights the sexual facets. This is the perspective most often associated with romantic love, and it assumes that romantic love is interlaced with sexual desire . . . which is not necessarily true. We'll come back to this when we discuss pair bonding and marriage. It is the eros aspect of love that underlies the common slang phrase "to make love" in reference to sexual activity. But love is much more than sex.

"Agape" initially implied a form of altruistic, giving, and all-encompassing love that is not sexual or even necessarily physical. It may have arisen in association with ancestor worship and the establishment of communities' new engagements with place and sacred spaces. Seen in this way, agape may be linked to moral codes and could even be seen as a form of transcendent experience (the "giving" of oneself beyond solely the material experience). This pattern of human engagement appears in a range of practices and belief systems across many societies. A version of agape, adopted and developed through the Christian philosophical tradition, has come to mean the sort of love that the Abrahamic God has for persons and for creation, as well as those persons' love for God. By extension, for Christians, it also reflects humans' love for each other, specifically in the context of their relation to God.[6] More recently, agape has been applied outside of a religious context and envisioned as a form of political and eco-philosophy, an unconditional, universal, brotherly, or altruistic love for humankind and the planet.

"Philia" is most often represented as a deep and affectionate regard, an intense allegiance and devotion—what we might call profound friendship—toward family members, friends, or associates; one's ethnic group, country, or nation; even an ideology. In this sense the notion of philia reflects the total commitment of the body and emotions in the act of being drawn toward another. Usually this is represented as a positive devotion. Philia is a common prefix or suffix in English, used to denote a specific attraction or devotion. Its opposite, "phobia" or fear, usually casts philia as a positive reality. However, philia can also reflect attachments that are dysfunctional and damaging (necrophilia, pedophilia). It can be a "bad" kind of love.

All three categories (eros, agape, philia) are often intertwined in popular concepts of love. They share the central theme of a deep and emotionally charged commitment to another.

Much contemporary scholarship on love makes an effort to distinguish between very strong affections and more regular affinities— "loving" as opposed to "liking." Many experts define "to love" as dedicating and investing oneself more intensely and more thoroughly than "to like." Philosopher Martha Nussbaum explains the difference by telling us that being in love with another is akin to embracing one system of values over another— not just an affinity but a cognitive and emotional commitment.[7] If we think back to chapter 6, we can envision the systems and processes by which the human organism and human societies coordinate neurobiologies, hormone systems, and community practices to facilitate this kind of commitment.

Many scholars and most of the public share Nussbaum's approach. Rather than dissecting the concept of love and placing strict boundaries on it, they see it as reflecting a particular state: a commitment, a devotion, an investment of the deepest meaning, centered around a profound capacity and commitment of care.[8] Being "in love" is a particular state of being that combines physical, social, and even transcendent aspects.

Contemporary philosophical scholarship seeks to classify love into specific contexts for different manners of analyses. Four common classifications of love are: love as union (the desire to form a specific union such as romantic or familial), love as robust concern (a commitment, dedication, emotional investment, or compassion), love as valuing (either seeing the "true" value of an individual or idea, or the bestowing onto that individual or idea the highest value), and love as a multifaceted emotion that may contain all of the above qualities.[9]

While there are clearly common threads among all of these uses of "love," there is no uniform definition. Love is not bounded by the material world or delimited by strict definitional boundaries. Instead, it resides simultaneously in the material, the practical, and the transcendent. If belief is the ability to draw on our range of cognitive and social resources, histories, and experiences in order to "see" and feel and know "something," and to utterly invest wholly and authentically in it such that it is one's reality, then love certainly requires belief to exist.

But when most people think or talk about love, they are not much concerned with scholarly discourse. Countless works of poetry and prose, songs, religious sermons, political statements, and entire genres of literature focusing on love continue to be produced. Many popular movies worldwide have love as a theme. To varying degrees, the forms of love described above play centrally in the beliefs and daily practices of every human society. Even using the most superficial index of global popularity, a Google search, "love" shows up in more than 14 billion results,[10] outperforming "death," "sex," and "hate." Love is central to people's lives and beliefs.

Why, in an evolutionary sense, are humans capable of this particular kind of deep attachment and commitment?

The Evolution of Attachment

If love is felt deeply and leaves marks on the body, then understanding something about its biology is important as the baseline for a discussion of it. What is going on in the body when humans feel so strongly toward one another? Why are these feelings so powerful? These bodily processes are often at the heart of what many people mean when they talk about love—especially when they talk about the evolution of love.

Popular discourse often positions love as the center of romantic relationships and familial bonds, and this approach has some merit. But it is not the whole, or even the most accurate, picture. We know that humans form pair bonds, are frequently in very strong sexual and social relationships, and that family members and close friends often feel close ties to one another.[11] Yet these behaviors and characteristics do not exactly mean what most people think they do.

We know that humans have evolved a system that uses social and physical interactions, hormones, and the brain to prime the body to feel closer and more attached to other individuals. In the increasingly socially complex and information-rich groups that genus *Homo* evolved over the past 2 million years, this attachment was critical to our ability to get along. In the most basic sense, the system that humans deployed to create and maintain such intense bonds is the same system of social connections that all mammals share. It evolved in animals that give live birth and have to care intensively for their young, a group that includes the primates. We know that a suite of hormones and neurotransmitters—including oxytocin, vasopressin, prolactin, testosterone, and dopamine—are involved in developing and maintaining physiological bonds between mothers and infants in all mammals, and between fathers and infants in some mammals (especially in humans).[12] We also know that across our evolutionary history, these bonds have become extremely important in the extended and intensive nurturing and development of young humans[13] and in the specifics of how we behave and believe.

This psychological, behavioral, and biological system of physiological-social bonding has also been co-opted to function in more or less the same way in interactions outside the care of offspring. Our psychoneuroendocrine system is triggered by physical touch, intense social interactions in physical proximity or contact, and the overall patterns of intensive social interactions between individuals in groups. Anthropologist Walter Goldschmidt noted that humans across the world all seek out these strong bonding interactions; he called this pattern "affect hunger"[14] and suggested[15] that the basic system that bonds mammalian mothers to their infants has been expanded in the human species into a social and physiological bonding system between individuals of all ages, sexes, and genders.[16] The particularly strong drive of affect hunger, he argues, enables humans to form and experience types of social bonds across a wider range of individuals, and with a different type of quality and structure, than most other animals are capable of. Many researchers argue that these bonds have enabled humans to fare better than almost any other mammal on the planet.[17]

So one answer to "What is love?" is that it is a function of our biology and behavior. One key aspect of what we call "love" emerges from the intense possibilities for bonding that emerged during the development of the genus *Homo* over the last 2 million years. The "affect hunger" found in so many mammals has been reshaped into a particular ability to form many kinds of extremely strong social bonds across diverse targets—and this process has become a constitutive element of the human niche. The deep connections between our psychoneuroendocrine system and a particularly human cultural niche form the groundwork for why we bond so strongly.

We can see that this aspect of love matches up with the philosophical notions of agape or philia. We can also state that expressions of love between parents and offspring, between siblings or other family members, between good friends, and between romantic pairs generally have the same evolutionary base.

Much of the public, however, and many researchers[18] want to see love as something distinct from the deep attachments that characterize humans and some other animals. They want it to be about unique relations between romantic pairs (eros). Culturally, especially in the Western world, we often see romantic love as separate from familial love or friendship. Yet aside from slightly different hormone levels and other physiological responses related to sexual activity, romantic love is not biologically different from any other kind.[19] Of course there are substantive variations in the psychological and social details of what people perceive, experience, and believe about romantic love. The idea that romantic love is distinct from other deep attachments is a product of cultural beliefs and worldviews, not our biology.[20] In societies where it is common, "romantic love" is generally assumed to be about sex, reproduction, attraction, and sometimes marriage. But what such assumptions are focusing on is not actually romantic love but rather a particular relationship between humans that scientific researchers call a "pair bond."

THE PAIR BOND

In the basic biological sense, a pair bond is a special, predictable relationship between two adult animals. When researchers look at pair bonds in humans and some other mammals, especially primates, they focus on a relationship between two individuals (usually but not always a male and a female) that involves tight social connections and a sexual affiliation, typically including mating and the raising of young.[21] It is often stated that this pair bond is the basis of human society and a major early development in human evolutionary history.[22] But this is not quite correct.

Evolutionary history suggests that pair bonds are not necessarily linked to procreation, or to the nuclear family in humans and some of the other primates.[23] Substantial evidence shows that humans seek both social and sexual pair bonds,[24] but these do not *necessarily* involve sex, marriage, monogamy, or heterosexuality.[25] Marriage is not equal to evolutionary or physiological pair bonds, and the nuclear family is not the basal unit of human social organization.[26]

In a biological and evolutionary sense, there are two types of pair bonds: social and sexual.[27] The social pair bond is a strong behavioral and psychological relationship between two individuals that is measurably different, in physiological and emotional terms, from general friendships or other relationships. The sexual pair bond is a behavioral and physiological bond between two individuals with a strong component of sexual attraction. The participants in the sexual pair bond would rather have sex with each other than with other individuals. This does not imply exclusivity of either attraction or sexual activity, only a strong preference for a specific partner. In humans and some other mammals, pair bonds are developed via social interactions combined with the physiological activity of neurotransmitters and hormones such as oxytocin, vasopressin, dopamine, and corticosterone.[28]

Pair bonds in humans, both social and sexual, are part of the social networks that emerged as central patterns over the last 2 million years as genus *Homo* ramped up its capacities for cooperation and social complexity.[29] Pair bonds can involve multiple kinds of sexual relationships, in many societies involving what we could call "romantic attachments." Importantly, our sexual pair bonding, like our sexual activity, is not focused only on reproduction[30] and can play broad and important roles in human social relations. These pair bonds are not the same as marriage and not necessarily connected to monogamy or any particular sexual orientation. In addition to the sexual pair bonds, social pair bonding across genders and ages is extremely common in humans, probably more so than in most other species.[31] We can have social pair bonds with relatives and close friends, and with individuals of the same or different sexes, or the same age or different ages.

Pair bonds and the patterns of behavior and physiology they entail are central parts of the human experience. Our capacity for creating and maintaining these remarkably deep attachments establishes a critical infrastructure enabling humans to believe in the particular experience we call "love."

WHAT ABOUT MARRIAGE?

A majority of societies today identify marriage as an ideal goal for humans and have a suite of institutions, laws, and beliefs about how and why it should be undertaken. But a society's understanding of marriage may or may not include something we might call "love" as a necessary component. Marriage itself, as a specific structural and legal process in human societies,[32] is not deeply embedded in, or even emerging from, human evolutionary patterns. Nor is it necessarily connected to any of the facets of love we've discussed.

The view of marriage that now dominates the world's global religions and the cultures intertwined with them is, like those belief systems themselves, a recent occurrence in human history. Even more recent is the idea of a connection between marriage and romantic love. This idea gained prominence in much of Europe during the sixteenth century and rapidly spread across the Western world and then around much of the globe.[33] The view that marriage is the ultimate outcome for a couple in love is thus extremely recent. Before the last few centuries, and even in many societies today, there was no necessary connection between romantic love and marriage, and the obligations and expectations of marriage vary quite broadly.[34]

But when many scholars opine about the subject, they persistently confuse "marriage" with "pair bond."

Most social scientists and historians agree that marriage developed as a social system for legitimizing reproduction, determining the inheritance of property, and regulating sexual activity, and that it has only recently been conceived of as the culturally sanctioned outcome of romantic love.[35] Marriage, today, is an important way for many societies to officially recognize and sanction sexual pair bonds and any offspring that result. While human sexual pair bonds naturally occur both heterosexually and homosexually, recent (in an evolutionary sense) social and political histories of many societies have created contexts where heterosexual pair bonds are sanctioned and homosexual ones are not (although that is changing in some places[36]).

Are all married couples sexually or even socially pair bonded? Given the enormous variation in why and how people marry, probably not. But there is little research on these questions, and scant data. We do know, however, that the capacity to form these bonds, regardless of their association with marriage, has had a significant role in the human niche, and that such bonds almost always play a central role when humans formulate ideas about love.

DID COMPASSION EVOLVE INTO LOVE?

In 2018 an orca female named J-35 (also called Tahlequah) gave birth to a calf who soon died. J-35 continued to swim in the waters off San Juan Island, carrying her dead calf for 17 days. This enormous expenditure of energy held no survival benefit for the orca. It was an intense display of attachment and grief. J-35 showed significant compassion for her dead offspring.[37] Many called it love.

Compassion manifests in many forms across the animal kingdom. In mammals, the bonds between mother and offspring are intense and can result in displays of a deep emotional investment that defy purely functional, survival-based explanations. The bonds between members of tightly knit social groups of primates, wolves, hyenas, orcas, or elephants can appear as intense as those among humans. Giving aid and assistance, and providing deep social support to closely bonded others, are features of many mammals' lives. But in most mammals,[38] this caring does not regularly extend beyond mother-offspring pairs, beyond close pair bonds or friendship, or with regularity to the aged, injured, and infirm in the social group. In humans, compassion is more diffuse and ubiquitous, and it appears to be a significant component of the human niche.

Think about the complexities in forming and maintaining the kinds of communities our ancestors did over the last few million years. There must have been countless times when individuals fell ill, were injured, or grew weak with age. In most other animals, the injured or ill member of the group frequently becomes isolated, is sometimes even attacked, and slowly moves off and disappears (elephants might be an intriguing exception to this pattern). One of the amazing transitions in the human niche was the emergence of consistent caring behavior that kept at least some of the injured, sick, and aged alive and part of the community. We know that very early on, our ancestors expanded the psychoneuroendocrine basis of the mother-infant bond and spread it to a broader care capacity.[39] The ability to care for our children developed into an ability to care for others unrivaled in any other species.

Starting as early as 1.8 million years ago, we see evidence of the sharing of meat and other important food sources[40] to an extent that far exceeds what we see in most other social mammals. Beginning around the same time, we also see males and females, young and old, caring for children. Both of these behavioral traits became core aspects of the human niche. But there is substantive evidence that our ancestors took this compassion a few steps further. Work by archaeologist Penny Spikins and her colleagues reveals evidence of extended compassion.[41]

At the 1.8-million-year-old site of Dmanisi in Georgia,[42] one of the aged inhabitants had lost all but one tooth many years before he died. Given his age and likely frailty it is very possible that others in the group provided food for him—perhaps mashing or chewing it for him. Another example comes from a 1.5-million-year-old site in Kenya where the remains of a female *Homo erectus* show evidence that she likely suffered from hypervitaminosis A,[43] a disease caused by too much vitamin A in the diet. This disorder can cause problems with bone density and damaging bone growths, both of which are present in this fossil. Hypervitaminosis A can take a long time to manifest in the bones, and her fossil shows she had it full blown. The disease had clearly been developing for several months, during which she would have suffered from nausea, headaches, stomachaches, dizziness, blurred vision, reduced muscle capacity, and fainting. These would have seriously limited her ability to contribute to the group or even fend for herself, but she survived. It's obvious she was cared for extensively.

At another site, Sima de los Huesos in Spain[44] (the same site where there appears to be something like burial), there is evidence of a child, likely around eight years old, with a birth defect called "lambdoid single suture craniosynostosis," in which the bones of the skull fuse prematurely. This condition severely affects brain growth, leading to significant mental disabilities, locomotor challenges, and disfigurement of the face and head. This child lived at least five years or more with this disorder, looking and acting very differently from others and needing a lot of assistance and care—which he clearly received.

These few cases might not seem like much, but when we consider the entire collection of fossils we have from these time periods, the existence of all such examples tells us that compassion for others was, if not widespread, at least perhaps a contributing factor in the development of the human niche during this period.[45] Its expression gradually extended across communities and populations in the form of increasingly regular and extensive investments in caring for offspring, for ill individuals, and for each other.[46]

That was only the beginning. More recently, we have extended this capacity for deep compassion beyond our communities, beyond even species boundaries, to strangers, animals, objects, and even abstract concepts (like "God" or "nation"). If believing means drawing on our cognitive and social resources, our histories and experiences, and combining them with our imagination to devote ourselves to an idea or concept, then the capacity for compassion certainly plays a central role in such a system. One can easily see that compassion is a central feature in a wide array of human beliefs, particularly for the idea of a deep and devoted compassion for others—what we often call "love" (as in philia or agape).

But there is a flip side to intensely cooperative and compassionate communities—an opposite to love of this kind. Sometimes the more tightly we are bound within our community, the more leery we are of others. Our compassion for our own group can be harnessed to foster fear and hatred of others. For most of human history, low population densities kept groups largely separate. When they did encounter one another, it was usually beneficial to spend time together, collaborate, and even exchange members. These early communities were small enough that they did not usually have to compete with each other for food or space, and the mixing of members between groups was probably healthy socially and biologically. But as human communities grew, groups coalesced, developing stronger senses of identity, property, inequality, and association with specific places due to agriculture, storage, and other deep investments in land. These patterns enhanced the feelings of "us" and "them." At this point, evidence for the flip side of compassion—cruelty and conflict—becomes more common in the archaeological record. Love for one's own group is increasingly likely to be paired with hate for other groups. Belief in the possibility and capacity for love also offers the possibility of hate, and plays a central role in many societal conflicts.

WHY WE BELIEVE IN LOVE

Most of us believe that something like love, broadly construed, is deeply embedded in humanity. The Beatles told us it is all you need. And while it is not *all* we need, it is something most of us can experience and perhaps all of us desire. But love is not a single thing. It's a word that we use to gloss over the amazingly diverse, messy realities of human relationships, bonds, investments, and commitments. Being human is always much more interesting than whatever labels we apply to it.

In the preface to this book we saw that philosopher Søren Kierkegaard described believing as an act of being wholly and completely in love with a concept, an experience, or a knowledge.[47] Most of us, hearing that comment, seamlessly understand Kierkegaard's analogy between believing and being in love. It immediately evokes both the depth and the dedication of one's commitment. For most people, the power of the word "love" cannot be overestimated. Belief and love are intertwined. To make any of the physiological and philosophical processes that we've reviewed emerge and manifest as love, humans have to believe in some, or many, forms of it.

The Dalai Lama tells us that "the need for love lies at the very foundation of human existence. It results from the profound interdependence we all share with one another."[48] He's right. The deep human commitment to cooperative sociality, our intense tendency to pair bond, our communal caring for our young, and the immense capacity for compassion that is part of the human niche set an evolutionary baseline for contemporary humanity—a baseline that enables the development and experience of love in many forms. But this baseline is a capacity, a potential, not a blueprint or a destiny.

Many other animals also have incredible capacities for compassion. Many mammals exhibit deep caring for their offspring and their group mates. In humans, however, a suite of physiological and cognitive capacities have developed, enabling the emergence of patterns that we associate with eros, agape, and philia—categories into which fall the myriad of philosophical, psychological, bodily, and transcendent experiences we call human love. The particulars of how the body initiates, maintains, and alters the psychoneuroendocrine patterns associated with love are contingent on our social and historical realities. How our cognitive processes conceive of and perceive love, and how we in turn respond to these processes, are at least partially contingent on place, time, society, age, sex, gender, group, and any number of other factors. How we engage in love is an outcome of the ways the dynamic processes of human culture structure and direct our perceptions, expectations, and responses.

Humans pair bond, exhibit intense compassion and hatred, feel deep sexual attraction and infatuation, and wholly commit to ideologies, theologies, and philosophies. But our specific experiences of love are contingent on our particular beliefs about it. The human niche facilitates love, but human beliefs about love structure how our physiologies and minds react to experiencing it. This is why, despite the significant commonalities across all the patterns we call love, we experience it in such a great diversity of ways.

Beliefs and belief systems, as central components of human culture, permeate and shape human neurobiologies, bodies, and ecologies and act as dynamic agents in our evolution. Our capacity to believe emerges from evolutionary processes and is intrinsically tied to our abilities to imagine, to be creative, to hope and dream, and to infuse the world with meanings. This enables us to love.

CHAPTER 11

Does Belief Matter?

OES BELIEF MATTER? Clearly the general answer is yes. But I pose this question not in a general sense but with a specific focus on our immediate future. How does belief matter right now?

We know the capacity for belief, and the countless ways in which we deploy that capacity, is central to human evolution and has played a key role in our success as a species. But now that very success has brought us face-to-face with potentially catastrophic repercussions for much of humanity, other species, and the world.

Humans have become a dominant force—perhaps *the* dominant force—in global ecosystems. With that role come ethical and practical responsibilities hitherto unknown to humanity. Can belief matter for the benefit of humans, and others, in the twenty-first century? The answer is complex and not entirely clear. But my sense is that if we cannot more positively engage with *belief* to increase the good in the world, we will be in serious trouble.

In this final chapter I connect the key data and themes of the previous chapters to offer a suite of possibilities for the future. I highlight a few of the most dangerous contemporary patterns of belief. I also argue that, given the knowledge covered in this book, we can work to make belief matter in beneficial ways, but that a better path is neither easy nor guaranteed. The perspective I present is not (I hope) one of naïve hopefulness or idealistic dogmatism, but one that is influenced by the data and results from nearly twenty years of research into this topic. My years of study have made me a cautious optimist. Given what we know about human histories and capacities, I believe that the human niche, with its central role for belief, continues to hold great promise.

BELIEF

Centuries of scientific research, and millennia of philosophical and theological discussions, demonstrate that humans have an enormous capacity for "shared reality," a mutuality of cognition, experience, and perception. We are never alone, even inside our own minds. Shared reality is the basic infrastructure of human communication and belief systems, and it begins at the very start of our lives. Psychologist Ulf Liszkowski has shown that, even early in infancy, "interactions entailing different perspectives and minds emerge . . . based on prior interactional experiences . . . the emergence of shared reference is itself mediated by interaction and caregivers' assistance in goal directed activities. . . . Social understanding and communicative skills develop through a continuous process of mutual shaping between social interaction and cognition."[1] Shared reality for humans starts with our first breaths, our immediate immersion into community, and is more than the ability to think ourselves into the inner worlds of others. It is also the obsessive habit of thinking others into our inner world.

In the preface I cited Terry Eagleton when, nodding to Kierkegaard, he described believing as an act of being wholly and completely in love with a concept, an experience, a knowledge. I said that believing is a commitment, a devotion to possibilities, that need not be rooted in daily material reality. Believing can be committing to an unknown or a perception, or it can be a certainty that cannot be seen or measured. Belief that we can become things we are not yet is one of the fundamental ingredients in long-term motivation for self-transformation. Belief changes our motivations and shapes our behavior, and it does so in groups, going beyond individual psychology and generating deep solidarity in ways that have significant consequences.

Belief is a capacity that evolved in the human lineage over the past 2 million years.

AND IT IS NOT ALWAYS GOOD

My descriptions of the evolution and functions of the human capacity for belief have largely focused on the positive and the hopeful. That said, from the definition I've developed here, one can also see the dangers inherent in such a capacity.

To illustrate this I'll touch on three examples of how belief matters for the worse. The first example centers on misleading beliefs about the world and our relationships with it that have created lethal threats to our species and most others. The second example is about how we have deployed political and economic beliefs to create a system of such profound structural inequity that we risk losing any hope of shared justice. The third example is the abuse of belief systems via specific kinds of fundamentalism, including one specific thread of belief in "scientism"—a process that hampers our ability to work together and move humanity toward better outcomes.

Example 1. Today, many humans believe that the world should be exploited for our benefit. Many societies have limitations and exceptions to this notion, and some even practice sustainable use of plants, animals, and land.[2] But almost every society, religion, and political and economic entity believes it has a right, even an obligation, to use the world as it does. In the twenty-first century, market-driven economic systems have become dominant, and populations have increased at a record pace.[3] The resulting explosion in the extraction of resources has reached such a scale and impact that the only way humans could possibly continue at this pace is if we believed the earth can sustain us regardless of what we do to it. This belief is materially incorrect.

Societies at the forefront of the extraction and control of resources act as if they believe in a right to the planet, that humans own it.[34] Those of us in such societies often call the earth "our planet," but we rarely share our ownership claim with the millions of other species that also live here. Many of us believe that our creativity, technology, and innovation will pull us through any global crisis. These beliefs are reinforced by numerous religious, economic, and political ideologies, and they are deeply flawed.

Will Steffen and his colleagues tell us that "the human imprint on the global environment has now become so large and active that it rivals some of the great forces of nature in its impact on the functioning of the Earth system."[5] In November 2017 the journal *Bioscience* published a letter signed by over 16,000 scientists from 184 countries titled "World Scientists' Warning to Humanity: A Second Notice."[6] It identified dangerous global trends—species extinction; ozone depletion; carbon dioxide emissions; oceanic dead zones; temperature increases; loss of forest, freshwater, and saltwater resources; and human population growth—all of which portend a global catastrophe. Also in 2017 an article published in *ScienceAdvances* focusing specifically on our close cousins, the other primates, surveyed every major extinction threat, population loss, and habitat crisis and determined that in each case, *Homo sapiens* is 100 percent of the cause.[7] These studies and thousands of others offer direct scientific evidence that contemporary practices have amazingly damaging global effects. However, while some efforts take place to counter the damage and change national and corporate behavior, simply offering data rarely sways beliefs at the level and impact necessary to effect lasting change.

The human commitment, investment, and devotion to drawing resources from our planet is not sustainable at our current pace and population. Particular beliefs about ourselves and our relations to the world have brought about a pattern of climate change and ecological collapse that may have created a material abyss that's too deep for humanity to escape.

Example 2. In addition to the ecological abyss, we have believed ourselves into a crater of structural inequity across social, economic, racial, and gender divisions so deep that it appears ready to entomb us. This is extremely true in the United States, where economic inequality, racial animosity, and xenophobic panic have divided the nation and done violence to bodies, lives, and communities.[8] For example, there are consistent racial gaps in income, infant mortality, education, and incarceration, but little national will to address them. The belief in race as a "natural" division of humanity and of racial differences as emerging from evolutionary (biological) histories is ingrained in the U.S. psyche, and it is incorrect. We know that racial differences are not biologically driven (races are not biological units) or part of a human nature.[9] We have ample data that demonstrate that that races and racism, as social, historical, and political realities, are devastatingly real with a myriad of negative outcomes for society.[10] Again, the data themselves are not enough to sway beliefs for a majority of people.

In the United States, there are also substantive wage gaps in gender undergirded by beliefs about the natural capacities of men and women. There is a devotion to a particular form of market economy that has resulted in working-class purchasing power not growing in forty years and the top 1 percent of income earners having forty times more income than the bottom 90 percent.[11] And beliefs about immigration are dividing the country more today than in the past seven decades.[12] Such patterns of division and inequality, driven by beliefs, are not uncommon around the world today.

The human pace of world-shaping via belief systems that assume a naturalness to racial, gendered, and xenophobic inequities has outstripped many individuals' direct experience with that shaping. Too few people have firsthand recognition of the immensity of the changes, and we seldom stop to reflect on what we are doing. Because of our increasingly global commitments to specific political and economic systems, all of them based on beliefs about how the world is and should be, economic and political power are extremely skewed. In recent centuries, humans have developed immensely complex societies and byzantine political and economic structures that generate disconnection between groups, creating prejudices that keep people apart. Class restrictions, economic discrimination, racism, xenophobia, misogyny, and fear divide societies in increasingly rigid ways in this first quarter of the twenty-first century. Global conflicts and inequality based on interpretations of religious and political doctrines have been present for at least 5,000 to 8,000 years, but the current scale of differential impact, unequal distribution of resources, and disempowerment is entirely at a new level.

Today 8.6 percent of the global population owns 85.6 percent of the global wealth. Eighty-two percent of new wealth generated in 2016 and 2017 went to the top 1 percent of income holders, and none to the world's poorest 50 percent. Nearly 56 percent of the global population makes between USD $2 and $10 a day; a CEO of one of the top-five global fashion brands earns in four days what a Bangladeshi garment worker earns in a lifetime. These income differences reflect more lasting and severe inequalities of power, access, and health that are crushing the hopes of billions of people.[13] We are believing ourselves deeper and deeper into potentially irreversible landscapes of injustice.

For humans, ecosystems are not just physical realities. Economic systems are not just about production, consumption, and distribution, and political systems are not just about governance. All are constructed and supported by *beliefs* and embodied in the behavior of those living within them. For humans, belief, hope, and practice are forever intertwined with ecologies, bodies, and landscapes. Our ideologies bear on our evolution, and some contemporary ideologies have especially toxic implications for our species and disastrous impacts on many others.

Example 3. For my final case study I turn to what, as a scientist, is particularly concerning to me: the fundamentalism in the deployment of "scientism" as a tool of authority and a weapon against other belief systems.

In my 2017 book, *The Creative Spark*,[14] I argued that the practices of scientific investigation on one hand, and of faith-based ritual and devotion on the other, have been mutual components in the human niche for far longer than most think—and that they can enrich one another. But changes in structural and political landscapes, combined with changes in the ways in which scientific and religious practice are expressed, have led to a particular set of conflicts between them, especially around the concepts of the development of organic life via evolutionary processes. This has culminated, over the past century, into what we can call the "scientism versus religious fundamentalism" debate. Both sides of the debate are abusing the human capacity for belief in a similar way.

Here I draw heavily on the work of anthropologist Jonathan Marks,[15] who has written extensively on the hubris of scientists and the frequent and erroneous assumption by many scientists that they are working in a value-free, materially rooted zone of inquiry.[16] Leaving the specifics of the scientific method aside, much of what scientists do is infused with the beliefs, perceptions, and experiences that they, as members of a particular society, class, race, gender, and so on, bring to everything they do. We know this about scientists not only because it's been documented extensively by many scholars researching this topic but also because recent studies demonstrate that the outcomes of scientific research, in both quality and innovation, are improved when diverse groups of scientists work in collaboration.[17] Diversity of perspective and individual experience reduces confirmation bias, improves a team's ability to solve problems, and promotes creativity. Scientism, the belief that people doing science are an exception to the human pattern of being enmeshed in belief systems, is incorrect.[18]

We know that scientists bring to their research a specific suite of beliefs, not just about the practice of science itself. This is not a surprise. Such behavior becomes problematic, however, when certain scientists, often prominent ones, present what they believe is a "scientific" position as an argument against other modes of seeking knowledge about the world,[19] specifically pitting "science" against religious belief.

Of course, every religious belief system contains aspects that are clearly mythical or factually erroneous. Religions have developed over time, and most have been shaped over multiple languages, texts, traditions, and societies: they have history, like all belief systems. For example, the earth is not young, and humanity did not literally spring from two parents living in the Garden of Eden. Of course, most people who follow the Abrahamic religions do not interpret those aspects of sacred texts literally. Such details are not the issue here. The real issue is the use of the belief system of scientism to deny the possibility of other belief systems—and to denigrate anyone engaging in them.

Prominent scientists, arguing from this position, have claimed that the basic notion of a deity is a delusion and that followers of religious faiths are "duped" into believing a series of lies.[20] The core argument is that only a belief about the world that stems from a specific interpretation of certain scientific results has validity—that "science" gets at the real world while other systems of belief offer only illusion. This stance is ignorant for many reasons, but most importantly, it abuses what we know about patterns and processes of human belief and fails to acknowledge the ability of belief systems to collaborate, integrate, exchange, and beneficially influence one another.

Fundamentalism is the strict adherence to certain scriptures, dogmas, or ideologies, combined with an extreme need to maintain in-group versus out-group distinctions. Fundamentalism demands purity of thought. In its stark beliefs about what is "real" and what is not, scientism is a form of fundamentalism—especially when it entails total antagonism to the possibility of transcendent experiences as part of the *real* of human experience.

Any fundamentalism—religious, scientific, political, or economic—is dangerous to humanity. By definition, it curtails the infrastructure of belief: our ability to imagine, experiment, innovate, and entangle with others. These are the very skills that have proven again and again to offer humankind the greatest opportunities for success. I use scientism as an example here because, as a scientist, I personally experience it and see the danger it presents to the good that science, and scientists, can do in the world. But all fundamentalism is an abuse of the human capacity to believe.

Belief Can Matter If We Try

We have concrete and irrefutable evidence that egalitarian harmony does not exist in the human world. There was no Garden of Eden in our past, nor is there one in our future. Nor should we want one. So many origin stories depict humans as isolated from the world, content in their ignorance of the dangers, injustices, and complexities outside their blissful state. But no one has actually lived in such a state. From the earliest glimmerings of humanity, our ancestors sought out, absorbed, and manipulated knowledge. They met the challenges of dealing with one another, and with the world, via innovative, creative, compassionate, and conflictual methods; manipulating stone; and developing new ways to interact, forage, build and maintain bonds, teach, learn and imagine, and eventually to believe.

Knowing why and how we believe is central to making belief matter for the better in the future. Among the more than 7.5 billion of us alive today, belief systems are complex, often in conflict, and frequently in flux. Humanity is too far down the road of social, political, and economic structuring to not have inequity and structural violence as constant components of our societies. That much is a given. But we have considerable say in how inequity and structures of violence are managed, counteracted, and ameliorated. If there is to be any possibility of effecting changes in how we distribute power, access, and justice in our societies, belief must play a central role. It is the commitment to a suite of beliefs, and the actions that result from those commitments, that can redirect our trajectory toward more sustainable and just outcomes.

Two decades ago, I began a collaborative project in Bali, Indonesia, in a village called Padangtegal (which some readers might know as Ubud). The project focused on the interfaces between a few groups of monkeys, a Hindu temple complex, a forest, and the local and foreign humans who passed through, visited, and prayed there. When we began, one of our main interlocutors was Pak Acin, who was the head of the village council and de facto manager of the temple complex and the forest that surrounded it. One day early in the study, as we walked through the forest, he showed me the challenges the village faced. The forest had become stunted, and new growth had ceased to take hold. Villagers, priests, and others put forward many reasons for this, but Pak Acin had a hunch on which he wanted my opinion. Both villagers and visitors had once used banana leaves to wrap food and other items, and then thrown the leaves on the forest floor when they were done with them. When plastic wrapping became easily available, they switched to using it—but kept their old disposal habits. Our surveys showed that after years of this practice, the tangle of plastic under the topsoil was preventing new growth. When I passed along my team's findings, he nodded in affirmation: this had been his hunch. I offered our team's assistance in extracting the plastic and adding topsoil but also explained that to maintain the benefits, he'd have to create a local law prohibiting the use of plastic in and around the temples. He laughed and told us that such a ban would not work, as what needs to change before action is belief. If no one sincerely believed in the edict, their behavior would be false and unsustainable. What he really needed to do, he told me, was to figure out which deity in the Balinese Hindu pantheon is most clearly the goddess of plastic, from whom they could draw insight to develop and disseminate the necessary local knowledge alongside the information from the scientists. People's behavior had to be enmeshed in a system of belief that drew on the imagination and creativity of the community.

Over the next year, our group of Indonesian and American scientists collaborated with the local village council, the Balinese priests and local workers at the temple, and the villagers to create a narrative about plastics, forests, temples, and monkeys. This narrative became a set of mutually agreed-on guidelines and eventually local custom. We did not literally invent a goddess of plastic, but we did think within local traditions and processes. Our collaboration focused on one of the key Balinese philosophical and theological themes—*Tri Hita Karena*,[21] the belief in and practice of harmony between peoples and environments. This helped reestablish specific caretaking relationships between the people using the temples and the forest and soils around them, identifying plastics and disposal practices as disrupting balance. The temple community worked to emphasize the role of the monkeys and the forest in both a ritual and mythological context, and their economic roles, encouraging a renewed desire in the locals and visitors to coexist and share the temple and forest space amicably and healthfully.[22] In following Pak Acin's guidance, we mutually developed a range of actions linked to local perceptions to make the critical information accessible, to make it matter, and to effect change via belief. Science alone found one answer to the community's problem, but it could not have persuaded people to do what was needed. Local beliefs and perceptions, their traditions, could move people to action, but without specific connections, informed by the science, they might not have found a lasting solution to the problem. Restoring the forest's health required engaging both belief systems.

This story illustrates how two of the most important current belief systems need to come together in order to make changes happen on a grand scale. We need both devotion to scientific investigation and commitment to deep faith in the potential of humanity. It matters little whether you think of these different systems as C. P. Snow's[23] two cultures ("science" and "humanities") or as "science" and "religion." The future, if we are to have any hope at all, depends on collaboration, mutual investment, and sincere integration between these arenas of belief.

Niche Constructors and Believers

When I was six I lost a tooth. I was told to place it under my pillow and the tooth fairy would come in the night and replace the tooth with a coin. I did not buy that story for a moment, so I resolved to stay awake all night and catch my parents in the act of swapping the coin for the tooth. I remember the hours in the dark waiting faithfully for any noise, any movement. At one point I reached under the pillow to check on the tooth, and to my amazement, there was no tooth—only a coin! The jolt of wonder and awe that went through me at that moment still makes me shiver. My whole notion of reality was cracked open, and I spent the rest of the night in a sort of transcendent fugue.

The truth, of course, is that I did fall asleep during my vigil, and my parents made the swap then. But I did not realize that at the time. I "knew" I had stayed awake and that the tooth fairy must have come. I suddenly believed in a world in which there were tooth fairies. That belief did not last, but the experience has shaped my science. The experience of the transcendent has never left me, because I am human, and believing is what we do.

I do not mean to imply that the diverse belief systems around the world, especially those that engage with faith traditions, are akin to a child believing in a tooth fairy. But the transcendent moment I felt was an infinitesimally small reflection of the ubiquitous human capacity for belief—a capacity that, in our current situation, must be recognized, appreciated, developed, and harnessed.

At the outset of this book I asserted that believing is a commitment, an investment, a devotion to novel ways of imagining or becoming—none of which need be rooted in material reality but all of which can be infused with hope. Belief, for better and worse, is a deeply and distinctly human process, neither "accident nor miracle" but the product and process of our lineage's history and the evolution of the human niche.

Philosopher Simone de Beauvoir once said, "One's life has value so long as one attributes value to the life of others, by means of love, friendship, indignation, and compassion." The major way we attribute value is via belief. And I hope that we can better direct our shared beliefs and values to imagine, hope for, and create a better future.

Acknowledgments

I THANK MY partner, Devi Snively, for her amazing insight on, assistance on, and critiques of various drafts of this book and for more than twenty-five years of examining life together. I owe a great debt of gratitude to Celia Deane-Drummond, Greg Downey, Barbara King, James X. Sullivan, and Andrew Whiten for their feedback on versions and sections of the book manuscript. And I am deeply grateful for the amazing editing and insightful commentaries provided by my editor at Yale University Press, Bill Frucht.

The initial drafts for this manuscript took shape during my stay in Edinburgh, where it was my great honor to deliver the Gifford Lectures in early 2018. I have immense gratitude to David Fergusson and Anna Conroy; the principal and vice chancellor of the University of Edinburgh; and the other members of the Gifford committee for their invitation to deliver the 2018 lecture series and for their wonderful and gracious hosting of me and my partner during the lectures.

Too many colleagues, friends, and family to name helped me think through aspects touched on in this book. However, I want to thank the following for their particularly generous discussions on these topics: Neil Arner, Marcus Baynes-Rock, Susan Blum, Oliver Davies, Celia Deane-Drummond, Greg Downey, Elizabeth Fuentes, Victor Fuentes, Lee Gettler, Rita and Walter Haake, Barbara Harvey, Eileen Hunt Botting, Wentzel van Huyssteen, Tim Ingold, Marc Kissel, Greg Kucich, Kevin Laland, Daniel Lende, James McKenna, Owen Murphy, Rahul Oka, Walter Rushton, William Storrar, Aku Visala, Judy Wofsy, and Melinda Zeder.

As always I am indebted to my terrific literary agent and friend, Melissa Flashman, for her faith in me and her knack for connecting me to the right places for my work. I thank Karen Olson at Yale Press for her assistance, and am grateful to Susan Arellano, of Templeton Press, for her support for this book and the series in which it appears. I thank Trish Vergilio of Templeton Press and Bob Land for the copyediting. Melissa Flamson and Deborah Nicholls assisted me with obtaining the permission for the images and art in the book, and I am very grateful for their help.

Finally, I thank Shelley, the wonderdog, for her cross-species care, companionship, and inspiration during the writing of this book.

Notes

PREFACE

1. Sherrington, "Man on His Nature."
2. https://www.google.com/search?q=beleif+dictionary&rlz=1C1GGRV_enUS751US751&oq=beleif+dictionary&aqs=chrome..69i57j0l5.3479j0j4&-sourceid=chrome&ie=UTF-8.
3. Wikipedia, "Belief."
4. University of Edinburgh, "Terry Eagleton–The God Debate."

CHAPTER 1

1. Fox-Keller, *Century of the Gene.*
2. Campbell et al., *Primates in Perspective.*
3. Whiten et al., *Culture Evolves*; Whiten, "Scope of Culture in Chimpanzees, Humans, and Ancestral Apes." This is in no way to deny that we see many complex social realities, friendships, deep relationships, and personality variants in a wide range of organisms—from corvids to parrots, from horses to lions, and on to a broad array of species.
4. See Foote, "Genome-Culture Coevolution."
5. Strier, *Primate Behavioral Ecology.*
6. Campbell et al., *Primates in Perspective*; Strier, *Primate Behavioral Ecology.*
7. See, for example, Strier, "What Does Variation in Primate Behavior Mean?"; Tomasello, *Natural History of Human Thinking*; Van Schaik, *Primate Origins of Human Nature*; King, *Information Continuum*; Strum, "Darwin's Monkey"; Whiten, "Scope of Culture in Chimpanzees, Humans, and Ancestral Apes."
8. Which in turn is critical in human's capacity for belief.
9. Wikipedia, "Crittercam."

10. Teleki, "They Are Us."
11. Konner, *Tangled Wing*, 435.
12. See details in Fuentes, *Biological Anthropology*, and Wood and Boyle, "Hominin Taxic Diversity."
13. See Wood, "Reconstructing Human Evolution."
14. See Wood and Boyle, "Hominin Taxic Diversity," and Fuentes, *Biological Anthropology*.
15. Fuentes, *Biological Anthropology*.
16. Johannson, "Lucy."
17. McPherron et al., "Evidence for Stone-Tool-Assisted Consumption."
18. Hovers, "Tools Go Back in Time"; Harmand et al., "3.3-Million-Year-Old Stone Tools from Lomekwi 3."
19. Antón, Potts, and Aiello, "Evolution of early *Homo*," and Wood and Boyle, "Hominin Taxic Diversity."

CHAPTER 2

1. See an accessible summary of a myriad of species concepts by biologist John Wilkins at "How Many Species Concepts Are There?" *The Guardian*, October 20, 2010; also see Wilkins, *Species*.
2. Ingold, *Perception of the Environment*.
3. Von Uexküll, *Foray into the Worlds of Animals and Humans*.
4. See Wake, Hadly, and Ackerly, "Biogeography"; and Hutchinson, "Concluding Remarks."
5. Fuentes, "Extended Evolutionary Synthesis"; Andersson, Törnberg, and Törnberg, "Evolutionary Developmental Approach to Cultural Evolution"; O'Brien and Laland, "Genes, Culture, and Agriculture."
6. Fuentes, "Integrative Anthropology and the Human Niche"; Fuentes, "Extended Evolutionary Synthesis"; Fuentes, "Human Niche."
7. Deacon, "On Human (Symbolic) Nature."
8. Fuentes and Wiessner, "Reintegrating Anthropology." S3.
9. Laland et al., "Does Evolutionary Theory Need a Rethink?"
10. See the Understanding Evolution, "Natural Selection," accessed May 7, 2109, for a clear and simple explanation for standard natural selection: https://evolution.berkeley.edu/evolibrary/article/evo_25.
11. Laland et al., "Does Evolutionary Theory Need a Rethink?"
12. Here is the original paper by Darwin and Wallace: "On the Tendency of Species to form Varieties; and on the Perpetuation of Varieties and Species by Natural Means of Selection. By Charles Darwin, Esq., F.R.S., F.L.S., & F.G.S., and Alfred Wallace, Esq. Communicated by Sir Charles Lyell, F.R.S.,

F.L.S., and J. D. Hooker, Esq., M.D., V.P.R.S., F.L.S, &c." (http://wallace
fund.info/content/1858-darwin-wallace-paper).

13. Antón, Potts, and Aiello, "Evolution of early *Homo*"; Wood and Boyle, "Hominin Taxic Diversity"; Wood, "Reconstructing Human Evolution."

14. McBrearty and Brooks, "Revolution That Wasn't"; Marean, "Evolutionary Anthropological Perspective"; Shea, "*Homo Sapiens*"; Sterelny, "From Hominins to Humans."

15. Hublin et al., "New Fossils."

16. Shea, "*Homo Sapiens*"; Sterelny, "From Hominins to Humans."

17. Kissel and Fuentes, "'Behavioral Modernity' as a Process."

18. Kissel and Fuentes, "'Behavioral Modernity' as a Process."

19. Ackermann, Mackay, and Arnold, "Hybrid Origin of 'Modern' Humans"; Scerri, E. et al., "Did Our Species Evolve in Subdivided Populations across Africa?"

20. Kissel and Fuentes, "'Behavioral Modernity' as a Process."

21. See Ackermann, Mackay, and Arnold, "Hybrid Origin of 'Modern' Humans"; Hammer et al., "Genetic Evidence for Archaic Admixture in Africa"; Hawk, "What Is the 'Braided Stream' Analogy for Human Evolution?"

22. Fuentes, "How Humans and Apes Are Different"; Brooks et al., "Long-Distance Stone Transport and Pigment Use"; Deino et al., "Chronology of the Acheulean to Middle Stone Age Transition in Eastern Africa"; Kuijt and Prentiss, "Niche Construction"; Smith and Zeder, "Onset of the Anthropocene"; Zeder, "Domestication as a Model System"; Mattison et al., "Evolution of Inequality."

23. Hunley, Cabana, and Long, "Apportionment of Human Diversity Revisited."

CHAPTER 3

1. Antón, Potts, and Aiello, "Evolution of early *Homo*"; Fuentes, *Creative Spark*; Gamble, Gowlett, and Dunbar, "Social Brain and the Shape of the Palaeolithic"; Foley, "Mosaic Evolution."

2. Fuentes, *Creative Spark*; Kissel and Fuentes, "'Behavioral Modernity' as a Process."

3. Laland et al., "Does Evolutionary Theory Need a Rethink?"

4. Grove and Coward, "Beyond the Tools"; Grove and Coward, "From Individual Neurons to Social Brains."

5. See summary in Fuentes, *Biological Anthropology*.

6. Potts, "Environmental and Behavioral Evidence."

7. Hiscock, "Learning in Lithic Landscapes"; Sanz, Call, and Boesch, "Tool Use in Animals"; Sterelny, *Evolved Apprentice*; Stout et al., "Cognitive

Demands of Lower Paleolithic Toolmaking"; Stout and Chaminade, "Stone Tools."

8. Hart and Sussman, *Man the Hunted.*

9. Fuentes, Wyczalkowski, and MacKinnon, "Niche Construction through Cooperation"; Antón and Snodgrass, "Origin and Evolution of Genus *Homo.*"

10. Aiello and Antón, "Human Biology and the Origins of *Homo*"; Ungar, Grine, and Teaford, "Diet in Early *Homo.*"

11. Archer et al., "Early Pleistocene Aquatic Resource Use in the Turkana Basin."

12. Antón, Potts, and Aiello, "Evolution of Early *Homo*"; Antón and Snodgrass, "Origin and Evolution of Genus *Homo.*"

13. Foley, "Mosaic Evolution"; Antón and Snodgrass, "Origin and Evolution of Genus *Homo*"; Kuzawa et al., "Metabolic Costs and Evolutionary Implications of Human Brain Development."

14. Hrdy, *Mothers and Others.*

15. Van Schaik, *Primate Origins of Human Nature*; Kuzawa and Bragg, "Plasticity in Human Life History Strategy"; Kramer and Otarola-Castillo, "When Mothers Need Others."

16. Gettler, "Becoming DADS"; Hrdy, *Mothers and Others.*

17. Fuentes, *Biological Anthropology.*

18. Here "meaning-making" reflects the capacity to create novel utterances, signs, materials, and actions such that they convey more than just the shape, color, or literal intent; they create a sense of something that is more than the material at hand. A figurine of a half human–half animal, a colored amulet, a red pigment intentionally smeared across one's body or face—all are early examples of how materials can be used to mean more than their physical characteristics.

19. Hiscock, "Learning in Lithic Landscapes"; Sterelny, *Evolved Apprentice*; Sterelny, "Paleolithic Reciprocation Crisis."

20. Stout et al., "Cognitive Demands"; Stout and Chaminade, "Stone Tools"; Morgan et al., "Experimental Evidence."

21. Arbib, "From Mirror Neurons to Complex Imitation."

22. Foley, "Mosaic Evolution"; Fuentes, *Creative Spark.*

23. A blend of technologies and bodies; see Haraway, "Cyborg Manifesto."

24. Brooks et al., "Long-Distance Stone Transport and Pigment Use."

25. Spikins, *How Compassion Made Us Human*; Spikins, Rutherford, and Needham, "From Homininity to Humanity."

26. Kissel and Kim, *Emergent Warfare.*

27. Wrangham and Carmody, "Human Adaptation to the Control of Fire"; Wiessner, "Embers of Society."

28. Kissel and Fuentes, "'Behavioral Modernity' as a Process"; Kissel and Fuentes, "Semiosis in the Pleistocene."

29. Deacon, "On Human (Symbolic) Nature."

30. Bloch, "Imagination from the Outside and from the Inside"; Bloch, "Why Religion Is Nothing Special but Is Central."

31. Montagu, *Human Revolution*, 2–3. Montagu used "man" instead of "humans," but here I update the quote, and given what we know about Montagu and his thoughts on gender and language, I am sure he would agree with the rationale for doing so.

CHAPTER 4

1. Fuentes, *Creative Spark*; Smith and Zeder, "Onset of the Anthropocene."

2. Hodder, *Religion, History, and Place.*

3. Steffen et al., "Anthropocene."

4. Fuentes, *Creative Spark*; Kuijt and Prentiss, "Niche Construction"; Smith and Zeder, "Onset of the Anthropocene"; Zeder, "Domestication as a Model System"; Mattison et al., "Evolution of Inequality"; Zeder, "Why Evolutionary Biology Needs Anthropology."

5. Unlike many common assumptions, there is no one locale for the origin of domestication (such as the Fertile Crescent of the Middle East). Domesticatory relationships between humans, plants, and animals begin to emerge in multiple human populations around the planet starting around 15,000 years ago or so.

6. Smith and Zeder, "Onset of the Anthropocene"; Zeder, "Domestication as a Model System"; Zeder, "Domestication of animals"; Zeder, "Why Evolutionary Biology Needs Anthropology."

7. Larson and Fuller, "Evolution of Animal Domestication"; Hunt and Rabett, "Holocene Landscape Intervention."

8. Zeder, "Domestication as a Model System"; Larson and Fuller, "Evolution of Animal Domestication"; Zeder, "Why Evolutionary Biology Needs Anthropology."

9. Larsen, "Biological Changes in Human Populations with Agriculture"; Zeder, "Domestication as a Model System"; Larson and Fuller, "Evolution of Animal Domestication"; Shipman, *Animal Connection*; Smith, "General Patterns of Niche Construction."

10. Fuentes, *Creative Spark.*

11. Maher et al., "Unique Human-Fox Burial"; Davis and Valla, "Evidence for Domestication of the Dog."

12. Olmert, *Made for Each Other*; Shipman, *Invaders.*

13. Olmert, *Made for Each Other*; Shipman, *Invaders*.
14. Trut, Oskina, and Kharlamova, "Animal Evolution during Domestication"; Dugatkin and Trut, *How to Tame a Fox*.
15. Larsen, "Biological Changes in Human Populations with Agriculture"; Flannery, "The Origins of Agriculture"; Fuller et al., "Convergent Evolution."
16. Callaway, "Domestication: Birth of Rice"; Fuller, "Domestication Process."
17. Fuller and Weisskopf, "Early Rice Project."
18. Callaway, "Domestication: Birth of Rice"; Fuller, "Domestication Process"; Huang et al., "Map of Rice Genome Variation."
19. See, for example, Sheehan et al., "Coevolution of Landesque Capital-Intensive Agriculture."
20. Wadley, "Recognizing Complex Cognition"; Sterelny, "Artifacts, Symbols, Thoughts."
21. Bar-Yosef, "Natufian Culture in the Levant"; Ullah, Kuijt, and Freemanc, "Toward a Theory of Punctuated Subsistence Change"; Kuijt, "What Do We Really Know about Food Storage, Surplus, and Feasting in Preagricultural Communities?"
22. Bowles and Choi, "Coevolution of Farming and Private Property during the Early Holocene."
23. Boehm, "What Makes Humans Economically Distinctive?"; Bowles and Gintis, *Cooperative Species*; Fry, *Human Potential for Peace*.
24. Boehm, *Hierarchy in the Forest*; Bowles and Gintis, *Cooperative Species*; Fry, *Human Potential for Peace*.
25. Bowles and Choi, "Coevolution of Farming and Private Property during the Early Holocene"; Sheehan et al., "Coevolution of Landesque Capital-Intensive Agriculture"; Graeber, *Debt*; Mattison, "Evolution of Inequality."
26. For basic concepts and a summary of details of sedentism in the Neolithic Middle East, see Zeder, "Neolithic Macro-(R)evolution."
27. Hodder, *Religion, History, and Place*.
28. This does not mean that humans did not move; they did, a lot. It is just that the majority of most populations stayed put for many generations. Clusters of populations left and wandered far away, and other clusters came in (humans genetics indicates lots of such movement and mating), but on average it would not be uncommon for significant percentages of populations to spend tens or hundreds of generations in more or less the same places.
29. Hodder, *Religion, History, and Place*.
30. Clare, "Establishing Identities in the Proto-Neolithic." Also see a wide range of perspectives on this site in the article and commentary at Banning, "So Fair a House."
31. This was before we have any evidence of writing, and see chapter 8 for more about religion and how we understand it.

32. This is the general form of this concept taken from Durkheim, *Elementary Forms of Religious Life*, and Eliade, *Sacred and the Profane*.
33. Zeder, "Neolithic Macro-(R)evolution."
34. Zeder, "Neolithic Macro-(R)evolution"; Graeber, *Debt*.
35. See Zeder, "Neolithic Macro-(R)evolution"; Sheehan et al., "Coevolution of Landesque Capital-Intensive Agriculture."
36. See Fausto-Sterling, *Sex/Gender*; and Fine, *Delusions of Gender*.
37. Estalrrich and Rosas, "Division of Labor"; Kuhn and Stiner, "What's a Mother to Do?" See also Adovasio, JM, Soffer, O and Page, J (2007) The Invisible Sex: uncovering the true roles of women in prehistory. Smithsonian Books; Adovasio, Soffer, and Page, *Invisible Sex*.
38. Adovasio, Soffer, and Page, Invisible Sex.
39. Bentley, Goldberg, and Jasienska, "Fertility of Agricultural and Non-Agricultural Traditional Societies."
40. Macintosh, Pinhasi, and Stock, "Prehistoric Women's Manual Labor."
41. Graeber, *Debt*, and Box 1 from Mattison, "Evolution of Inequality."
42. Fry, *War, Peace, and Human Nature*; Ferguson, "War before History"; Ferguson, "Born to Live"; Fuentes, *Race*; Kissel and Kim, *Emergent Warfare*; Haas and Piscitelli, "Prehistory of Warfare"; Haas and Piscitelli, "Misled by Ethnography"; Ferguson, "Pinkers List"; Sala, "Lethal Interpersonal Violence in the Middle Pleistocene"; Gómez et al., "Phylogenetic Roots of Human Lethal Violence."
43. Pinker, *Better Angels of Our Nature*.
44. Kissel and Kim, "Emergence of Human Warfare."

CHAPTER 5

1. Williams, "Making Representations."
2. Williams, "Making Representations."
3. Montagu, *Human Revolution*.
4. Montagu, *Human Revolution*.
5. Brumann, "Writing for Culture."
6. Brumann, "Writing for Culture."
7. Ingold, "Art of Translation in a Continuous World."
8. Laland, *Darwin's Unfinished Symphony*.
9. Here I am bypassing a great deal of the debates and history in anthropology about the "culture" concept in favor of the mode of usage and functional definition that I outline in this chapter. However, for a recent, and very good, overview of some of the contemporary discourses on culture as a concept and a practice, see Rodseth, "Hegemonic Concepts of Culture."
10. Fuentes and Wiessner, "Reintegrating Anthropology."

11. Grove and Coward, "Beyond the Tools"; Fuentes, *Race*; Andersson, Törnberg, and Törnberg, "Evolutionary Developmental Approach to Cultural Evolution"; Downey and Lende, "Evolution and the Brain."

12. Taylor, *Primitive Culture*.

13. Kroeber and Kluckhohn, *Culture*.

14. Brumann, "Writing for Culture."

15. Whiten et al., *Culture Evolves*.

16. Ramsey, "Culture in Humans and Other Animals."

17. Whiten, "Scope of Culture in Chimpanzees, Humans, and Ancestral Apes"; Whiten et al., "Extension of Biology through Culture."

18. Hobaiter et al., "Social Network Analysis."

19. Sanz, Call, and Boesch, *Tool Use in Animals*. Thanks also to Dr. Andrew Whiten for the gardener analogy.

20. Wilson, "Chimpanzees, Warfare, and the Invention of Peace."

21. Foote, "Genome-Culture Coevolution."

22. I thank Professor Greg Downey for this particularly effective phrasing.

23. https://bizarro.com/2016/09/04/bush-voyeurs/.

24. Whiten et al., *Culture Evolves*.

25. Deacon, "On Human (Symbolic) Nature."

26. Whiten, "Scope of Culture in Chimpanzees, Humans, and Ancestral Apes."

27. Ingold, "To Human Is a Verb."

28. Bourdieu, *Outline of a Theory of Practice*.

29. http://www.mcescher.com/

30. Kroeber and Kluckhohn, Culture

31. Ackermann, Mackay, and Arnold, "Hybrid Origin of 'Modern' Humans"; Scerri, E. et al., "Did Our Species Evolve in Subdivided Populations across Africa?"

32. Kissel and Fuentes, "'Behavioral Modernity' as a Process." Of course, this statement can refer to all organisms, but here I am specifically referring to the process of members of the genus *Homo* becoming contemporary humans (see also chapters 2 and 3).

33. Sterelny, "Paleolithic Reciprocation Crisis"; Sterelny and Hiscock. "Symbols, Signals, and the Archaeological Record."

34. Kissel and Fuentes, "'Behavioral Modernity' as a Process."

35. Goldman, "Is Language Unique to Humans?"; Gots, "Humans Make Language, Language Makes Us Human"; Katz, "Noam Chomsky on Where Artificial Intelligence Went Wrong."

36. Barnard, *Social Anthropology and Human Origins*.

37. Martinez et al., "Auditory Capacities in Middle Pleistocene Humans from the Sierra de Atapuerca in Spain."

38. Staes et al., "FOXP2 Variation in Great Ape Populations."

39. See previous chapters for details.
40. Immanuel Kant. *Critique of Pure Reason*. Project Gutenberg.
41. Abraham, "Imaginative Mind"; Bloch, "Imagination from the Outside and from the Inside"; Fuentes, "Human Evolution."
42. Bloch, "Why Religion Is Nothing Special but Is Central."
43. Here I suggest reading a range of basic religious studies texts and works that cover broad introductions to philosophy and theology. The issue of transcendence is pervasive in these fields.
44. For the case that humans are distinctive in their abilities for abstract reason as rational and moral agents, see, for example, Macintyre, *Dependent Rational Animals*.

CHAPTER 6

1. Tomasello, *Natural History of Human Thinking*.
2. Boyd, Richerson, and Henrich, "Cultural Niche"; see also Patrick Clarkin's thoughts on this issue: https://kevishere.com/2014/06/10/developmental-plasticity-and-the-hard-wired-problem/.
3. Hare, "From Hominoid to Hominid Mind"; Tomasello, *Cultural Origins of Human Cognition*.
4. Andersson, Törnberg, and Törnberg, "Evolutionary Developmental Approach to Cultural Evolution"; Read, *How Culture Makes Us Human*; Richerson and Boyd, *Not by Genes Alone*.
5. Laland, *Darwin's Unfinished Symphony*; Ingold, *Perception of the Environment*; and Clarkin, "Developmental Plasticity."
6. Ingold, *Perception of the Environment*
7. Downey and Lende, "Evolution and the Brain."
8. See Han, *Sociocultural Brain*, for an excellent overview of a range of brain imaging and neurobiological studies that support this point.
9. See, for example, Gettler, "Applying Socioendocrinology to Evolutionary Models"; Lende and Downey, *Encultured Brain*; Rilling, "Neural and Hormonal Bases of Human Parental Care"; Van Anders, "Beyond Masculinity."
10. Han and Ma, "Culture-Behavior-Brain Loop Model"; Han, *Sociocultural Brain*; Sherwood and Gomez-Robles, "Brain Plasticity and Human Evolution."
11. Kuzawa et al., "Metabolic Costs and Evolutionary Implications of Human Brain Development"; Downey and Lende, "Evolution and the Brain."
12. Cognition is the acquisition, development, and use of knowledge, insight, and understanding
13. Barrett, "Why Brains Are Not Computers."
14. Meaning of semiotics: laden and creative.

15. Sterelny, "Artifacts, Symbols, Thoughts."
16. Laland, *Darwin's Unfinished Symphony*.
17. For Laland, these are cultural learning, intelligence, language, cooperation, and powers of computation.
18. Laland, *Darwin's Unfinished Symphony*.
19. Downey and Lende, "Evolution and the Brain."
20. Laland, *Darwin's Unfinished Symphony*.
21. Abraham, "Imaginative Mind"; Abraham, "Introduction."
22. Garderinck and Osvath, "Tripod Effect."
23. Carlén, "What Constitutes the Prefrontal Cortex?"
24. Whiten and van de Waal, "Social Learning."
25. Barton and Vendetti, "Rapid Evolution of the Cerebellum."
26. Sterelny, *Thought in a Hostile World*.
27. Donald, *Origins of the Modern Mind*.
28. Tomasello, *Natural History of Human Thinking*.
29. Han, *Sociocultural Brain*.
30. Luhrmann, "Towards an Anthropological Theory of Mind."
31. Also called "social constructs."
32. I am not asserting that all human beliefs, institutions, and actions are only understandable in their own context, or that there are no ways to differentiate those that are harmful vs. beneficent, altruistic vs. selfish, or malevolent vs. compassionate. Rather, I am stating that when someone believes something, it is real for that person. This is a critical assertion because, for humans, what is and what should be are critical components of every human belief system and are, in large part, both contextualized and contingent on where and how we develop and are embedded in who we are and how we become. I'll repeat: belief matters.
33. Peirce, *Logic of Interdisciplinarity*.

CHAPTER 8

1. Bellah, *Religion in Human Evolution*.
2. See chapters 3 and 4
3. Pew Research Center, "Executive Summary."
4. Rappaport, *Ritual and Religion in the Making of Humanity*.
5. Bellah, *Religion in Human Evolution*.
6. Donald, *Mind So Rare*.
7. This centrality and evolutionary relevance of religion to the human experience is elaborated on in the Anthropologist Melvin Konnor's most recent book *Believers: Faith in Human Nature*, 2019, Norton.
8. Geertz, "Religion as a Cultural System"; Geertz, *Thick Description*.

9. Durkheim, *Elementary Forms of Religious Life* (1995 ed.).
10. Crane, *Meaning of Belief.*
11. That is, believing that the mind and body are separate entities or dimensions of the human.
12. The mind is an independent nonmaterial entity (temporarily) inhabiting and controlling the body (a form of Cartesian dualism); see Koestler, *Ghost in the Machine.*
13. See chapter 4.
14. Kissel and Fuentes, "Semiosis in the Pleistocene."
15. Peirce, *Collected Papers of Charles Sanders Peirce*; Peirce, *Essential Peirce.*
16. Gvozdover, "Typology of Female Figurines of the Kostenki Paleolithic Culture"; McDermott, "Self-Representation in Upper Paleolithic Female Figurines"; Rice, "Prehistoric Venuses."
17. If interested, please see my previous summaries of the various proposals linking tools and ritual as it relates to this theme in Fuentes, *Creative Spark.*
18. An earthy clay with ferric oxide that can be yellow, red, or orange-brown in color.
19. Wadley, "Recognizing Complex Cognition."
20. For details of all of these examples and many more, see Kissel and Fuentes, "Database of Archaeological Evidence."
21. Kissel and Fuentes, "'Behavioral Modernity' as a Process"; Kissel and Fuentes, "Semiosis in the Pleistocene."
22. Carbonell and Mosquera, "Eemergence of a Symbolic Behaviour."
23. Dirks et al., "Geological and Taphonomic Evidence for Deliberate Body Disposal"; Context for the New Hominin Species."
24. Kissel and Fuentes, "'Behavioral Modernity' as a Process"; Scerri et al., "Did Our Species Evolve in Subdivided Populations across Africa?"
25. See chapter 3.
26. Wadley, "Recognizing Complex Cognition"; Kissel and Fuentes, "'Behavioral Modernity' as a Process"; Kissel and Fuentes, "Database of Archaeological Evidence."
27. Brooks et al., "Long-Distance Stone Transport and Pigment Use."
28. Rappaport, *Ritual and Religion in the Making of Humanity,* 1.
29. Wildman, *Science and Religious Anthropology.*
30. Van Huyssteen, "Lecture One."
31. Tomasello, *Natural History of Human Thinking.*
32. Atran, *In Gods We Trust*; Boyer and Bergstrom, "Evolutionary Perspectives on Religion"; Johnson and Bearing, *Hand of God*; Bering, *Belief Instinct*; Barrett, *Why Would Anyone Believe in God?*; Boyer, *Naturalness of Religious Ideas*; Sosis, "Adaptationist-Byproduct Debate on the Evolution of Religion."

33. Norenzayan, *Big Gods*; Norenzayan, "Does Religion Make People Moral?" Norenzayan, Henrich, and Slingerland, "Religious Prosociality."
34. Johnson and Bearing, *Hand of God.*
35. See details in chapters 3 and 4; and also see Whitehouse, et al. "Complex societies precede moralizing gods throughout world history," *Nature*, 2019.
36. Boyer and Bergstrom, "Evolutionary Perspectives on Religion."
37. Richerson and Boyd, *Not by Genes Alone.*
38. Pope, Russel, and Watson, "Biface Form and Structured Behavior in the Acheulean"; Kissel and Fuentes, "'Behavioral Modernity' as a Process"; Fuentes, "Human Evolution"; Fuentes, "Integrative Anthropology and the Human Niche."
39. Hiscock, "Learning in Lithic Landscapes"; Sterelny, *Evolved Apprentice.*
40. Rossano, "Ritual Behavior."
41. Sosis and Alcorta, "Signaling, Solidarity, and the Sacred."
42. King, *Evolving God.*
43. McNamara, *Neuroscience of Religious Experience.*
44. Van Huyssteen, *Alone in the World?*
45. Van Huyssteen, "Lecture Three."
46. Abraham, "Imaginative Mind"; Abraham, "Introduction."

CHAPTER 9

1. Here I am referring to a pattern of generalized belief about the economic nature of reality (primarily by those who are immersed in market economies). I am not delving into what has been termed "folk-economic beliefs" (e.g., the causes of the wealth of nations, the benefits or drawbacks of markets and international trade, the effects of regulation, and so on). Such beliefs are argued to be crucial in forming people's political beliefs and in shaping their reception of different policies. Recent work argues that these folk beliefs are rooted in a deeper evolutionary psychology that emerges across the Pleistocene. I do not agree with this deep evolutionary assertion, for reasons laid out in this chapter, but see the following article and associated commentaries for a recent overview of this perspective: Boyer and Peterson, "Folk-Economic Beliefs."
2. https://www.merriam-webster.com/dictionary/economy.
3. Luhby, "71% of the World's Population Lives on Less Than $10 a Day."
4. Here I am referring primarily to the beliefs and belief systems of people who grew up in towns and cities, and in regions and nations who rely on market economic systems, people who went to primary and secondary schools where they are taught that economics is a key to a successful human system,

and who are connected to, and dependent on, market economies locally and globally (this is around 88 percent of the humans on the planet).

5. Especially in the Euro-American tradition that forms the basis for much of global higher education over the past fifty years.

6. Weiss and Buchanan, *Mermaid's Tale*; Sussmna and Cloninger, *Origins of Altruism and Cooperation*; Gilbert, Sapp, and Tauber, "Symbiotic View of Life"; Sultan, *Organism and Environment.*

7. Remember, here we define an "economy" as an organized system of human activity involved in the production, consumption, exchange, and distribution of goods and services.

8. Money is a complex topic, but here I am using it in the general sense of an item or some form of record or notation that is accepted as payment for goods or services or repayment of debts and is the main component circulated in this manner in the market economy. However, money is far from that simple, as I discuss later in this chapter.

9. Wikipedia. "Market Economy."

10. A market economic system where private ownership controls the means of production of goods and their operation for profit.

11. In a command economy, all production and distribution are controlled by a central authority, and in a mixed economy, control is dispersed between public and private sources.

12. Hausman, "Philosophy of Economics."

13. https://www.aeaweb.org/resources/students/what-is-economics.

14. See, for example, Galor and Moav, "Natural Selection and the Origin of Economic Growth."

15. Fleischacker, "Adam Smith."

16. Smith, The Wealth of Nations.

17. I paraphrase from *Wealth of Nations.*

18. Smith, as with almost every male writer for much of history, uses only the male pronoun to represent all humans, even though his and most other philosophers' focus was indeed on men and not women, whom the philosophers considered secondary as contributors to society.

19. It is debatable that these beliefs undergird one of the world's largest economies: China.

20. Blaug, "Economics."

21. An increase in prices and decrease in the value of money—its valuation in regard to purchasing power.

22. Thanks to Professor James X. Sullivan for this insight.

23. Wolff, "Karl Marx."

24. Where the modes of production and distribution are primarily driven by private agents, not governments, and for profit.

25. Barker, "Happy Birthday, Karl Marx."
26. And some argue was never actually undertaken, as Marx had proposed.
27. Marx, *Poverty of Philosophy*.
28. Carrier and Miller, *Virtualism*.
29. Alós-Ferrer, "A Review Essay on Social Neuroscience," 261.
30. The study of behavior and interactions between organisms and the ecological pressures they face, in an evolutionary context.
31. https://www.nature.com/subjects/behavioural-ecology. See Davies, Krebs, and West, *Introduction to Behavioural Ecology*, for a classic and extensive overview.
32. Davies, Krebs, and West, *Introduction to Behavioural Ecology*.
33. Successful reproduction is the most commonly used measure of evolutionary fitness
34. Parker and Smith, "Optimality Theory in Evolutionary Biology."
35. Noë and Hammerstein, "Biological Markets."
36. Weiss and Buchanan, *Mermaid's Tale*; Sussmna and Cloninger, *Origins of Altruism and Cooperation*; Strier, *Primate Behavioral Ecology*; Strier, "What Does Variation in Primate Behavior Mean?"; Strum, "Darwin's Monkey."
37. Ha, "Cost-Benefit Analysis"; Fuentes, "Social Systems and Socioecology."
38. Bolduc and Cezilly, "Optimality Modelling in the Real World."
39. Kokko, "Give One Species the Task."
40. Again, I want to stress that the "we" here is mostly the developed world, people from societies and nations deeply committed to such economic systems. Many people from small-scale societies or those with less direct connections to market economies see the world differently. But there are fewer and fewer such people every year.
41. Henrich et al., *Foundations of Human Sociality*.
42. Small-scale societies are cohesive ethno-linguistic groups that are not fully immersed in industrial production and market economies or were not so until recently. Of the twelve there were foragers, horticulturalists, nomadic herding groups, and some sedentary agriculturalists.
43. See also Graeber, *Debt*, and Box 1 from Mattison, "Evolution of Inequality" for more ethnographic examples.
44. Boehm, *Hierarchy in the Forest*.
45. Wiessner, "Vines of Complexity."
46. And as we discussed in chapters 3 and 4, when we do have archaeological evidence of such roles, the patterns seem to emerge and develop in the material record of the past in the latest Pleistocene (the last 30,000 to 10,000 years ago).
47. Wiessner, "Vines of Complexity."
48. Boehm, "What Makes Humans Economically Distinctive?"

49. One that reflects a preponderance of shared goods, minimal economic role differentiation, and a lack of differential accumulation of or political control over goods.

50. Boehm, "What Makes Humans Economically Distinctive?"; Bowles and Gintis, *Cooperative Species*; Fry, *Human Potential for Peace*.

51. See chapter 4 again and Zeder, "Neolithic Macro-(R)evolution"; see also Hodder, *Religion, History, and Place*.

52. Mattison, "Evolution of Inequality," 184.

53. Brooks et al., "Long-Distance Stone Transport and Pigment Use"; Deino et al., "Chronology of the Acheulean to Middle Stone Age Transition in Eastern Africa."

54. Oka and Fuentes, "From Reciprocity to Trade."

55. Mauss, *Gift*.

56. Graeber, *Debt*.

57. Where values are assessed and the expectation of equitable trade is assumed.

58. See an excellent overview of this whole argument and the ethnographic data for it in Graeber, *Debt*.

59. Graeber, "Beads and Money"; Graeber, *Toward an Anthropological Theory of Value*; Maurer, "Anthropology of Money"; Hart, "Notes towards an Anthropology of Money."

60. Trinkaus and Buzhilova, "Diversity and Differential Disposal of the Dead at Sunghir."

61. Borgerhoff Mulder et al., "Intergenerational Wealth Transmission and the Dynamics of Inequality in Smallscale Societies"; Borgerhoff Mulder et al., "Pastoralism and Wealth Inequality"; Smith et al., "Wealth Transmission and Inequality among Hunter-Gatherers"; Kaplan, "Theory of Fertility and Parental Investment in Traditional and Modern Human Societies."

62. http://www.pewresearch.org/topics/income-inequality/; https://healthinequality.org/; Patten, "Racial, Gender Wage Gaps."

63. Marx, *The Philosophy of Poverty*

CHAPTER 10

1. In this chapter I am specifically not engaging, to any significant extent, with the debate about other animals and love. I have no doubt that many species display massive and deep attachment and caring for others and that deep feelings, yearning, and grief are found across the animal kingdom—and we review a few instances here. But "love" as I outline in this chapter is a human belief that includes these elements and also has a range of other facets tied specifically to the human niche and to human history. For a broader discussion of other animals, see King, *How Animals Grieve*.

2. Karandashev, "Cultural Perspective on Romantic Love"; Ryan and Jetha, *Sex at Dawn*; Goldschmidt, *Bridge to Humanity*; Hatfield and Rapson, *Love and Sex*; Fisher, *Why We Love*.

3. Karandashev, "Cultural Perspective on Romantic Love"; Ryan and Jetha, *Sex at Dawn*.

4. And to ideals or particular concepts or faiths and religious beliefs, as we discuss in the next chapter

5. Helm, "Love.".

6. In some Christian (and other religious) traditions there is also a fourth kind of love: *caritas*, the specific love of God (see Lewis, *Four Loves*). In such a view, agape is God's love for all but caritas is humans' love for God, and is more akin to friendship or even in some mystical traditions to desire or eros (without any sexual connotations). The classic Greek philosophy also allowed for self-love (*ohilatia*) and ludus (*playful love*).

7. Nussbaum,"Love and the Individual."

8. Helm, "Love, Identification, and the Emotions."

9. Stanford Encyclopedia of Philosophy, s.v. "Love," https://plato.stanford.edu/entries/love/.

10. As of December 15, 2018.

11. Fletcher et al., "Pair-Bonding"; see summaries in Fuentes, *Race*.

12. Rilling, "Neural and Hormonal Bases of Human Parental Care"; Van Anders, "Beyond Masculinity"; Gettler, "Becoming DADS."

13. Hrdy, *Mothers and Others*; Burkart, Hrdy, and van Schaik, "Cooperative Breeding and Human Cognitive Evolution."

14. Goldschmidt, *Bridge to Humanity*.

15. Such as the famous work of John Bowlby on attachment; Bowlby, J. (1969). *Attachment. Attachment and loss*: Vol. 1. *Loss*. New York: Basic Books.

16. Many researchers also point out that variants of this pattern, the co-opting of the mammalian mother-infant physiology for other inter-individual relations, show up in a number of other highly social mammals, such as primates, cetaceans, canids, etc. Humans are not unique in this, but we are quite distinctive in the specifics of how we deploy these patterns .

17. Hrdy, *Mothers and Others*; Richerson et al., "Cultural Group Selection Plays an Essential Role"; see also summaries in Fuentes, *Evolution of Human Behavior*.

18. Fisher, *Why We Love*; Fletcher et al., "Pair-Bonding."

19. See summaries in Fuentes, *Race*.

20. Barash and Lipton, *Myth of Monogamy*; Ryan and Jetha, *Sex at Dawn*.

21. Fuentes, "Patterns and Trends in Primate Pair Bonds"; Fuentes, "Hylobatid Communities"; Quinlan, "Human Pair-Bonds."

22. For recent versions of this approach, see Chapais, *Primeval Kinship*, and

Lovejoy, "Reexamining Human Origins in Light of *Ardipithecus ramidus*." For a broader overview, see Quinlan, "Human Pair-Bonds."

23. Fuentes, "Patterns and Trends in Primate Pair Bonds."

24. Quinlan, "Human Pair-Bonds"; Fletcher et al., "Pair-Bonding"; Gray and Campbell, "Human Male Testosterone."

25. Barash and Lipton, *Myth of Monogamy*; Fuentes, *Race*.

26. Ryan and Jetha, *Sex at Dawn*; see also Fuentes, *Race*; Barash and Lipton, *Myth of Monogamy*.

27. Fuentes, "Patterns and Trends in Primate Pair Bonds."

28. Ellison and Gray, *Endocrinology of Social Relationships*; Carter and Keverne, "Neurobiology of Social Affiliation and Pair Bonding"; Curtis and Wang, "Neurochemistry of Pair Bonding."

29. Quinlan, "Human Pair-Bonds"; Fuentes, "Re-Evaluating Primate Monogamy."

30. Ryan and Jetha, *Sex at Dawn*; Fuentes, *Creative Spark*; Fausto-Sterling, *Sexing the Body*; Martine, *Untrue*.

31. Other species do not have "gender" per se, but such bonds between and within sexes can be common in some highly social mammalian groups (in primates, cetaceans, elephants, viverids, etc.).

32. Squire, *I Don't*.

33. Squire, *I Don't*.

34. Waterston, "Marriage and Other Arrangements;" Squire, *I don't*.

35. Waterston, "Marriage and Other Arrangements."

36. Masci and DeSilver, "A Global Snapshot of Same-Sex Marriage."

37. Bekoff, "Make No Mistake, Orca Mom J-35 and Pod Mates are Grieving."

38. In elephants, caring for aged and injured individuals seems to be more common and more consistent than in most other mammals.

39. Hrdy, *Mothers and Others*; Burkart, Hrdy, and van Schaik, "Cooperative Breeding and Human Cognitive Evolution"; Tomasello et al., "Two Key Steps in the Evolution of Human Cooperation."

40. See chapter 2.

41. Spikins, *How Compassion Made Us Human*.

42. Lordkipanidze et al., "Complete Skull from Dmanisi, Georgia."

43. Walker, Zimmerman, and Leakey, "Possible Case of Hypervitaminosis A in *Homo erectus*."

44. Gracia et al., "Craniosynostosis in the Middle Pleistocene."

45. Spikins et al., "Calculated or Caring?"

46. Spikins, "Prehistoric Origins."

47. But we should note that for Kierkegaard the truest form of "love" was that between humans and the Christian God, the post-Greek agape. See Kierkegaard, *Works of Love*.

48. Gyatso, "Compassion and the Individual."

CHAPTER 11

1. Lizkowski, "Emergence of Shared Reference and Shared Minds in Infancy."
2. Hamilton and Grinevald, "Was the Anthropocene Anticipated?"; Steffen et al., "Anthropocene."
3. Including more than doubling of the world's population in the last fifty years" (http://www.worldometers.info/world-population/).
4. Despite not all of their populace believing in such extraction rates, the top ten GDP countries are good examples of this: FocusEconomics, "World's Top 10 Largest Economies."
5. Steffen et al., "Anthropocene," 614.
6. Ripple et al., "World Scientists' Warning to Humanity."
7. Estrada et al., "Impending Extinction Crisis of the World's Primates."
8. http://www.pewsocialtrends.org/, https://www.theguardian.com/us-news/2018/jun/06/growing-up-black-in-america-racism-education, http://www.pewresearch.org/topics/discrimination-and-prejudice/, http://www.pewresearch.org/topics/immigration-attitudes/2018/.
9. Hunley, Cabana, and Long, "Apportionment of Human Diversity Revisited"; Fuentes, *Race*; Bolnick, "Individual Ancestry Inference"; "ASHG Perspective"; http://www.understandingrace.com. See also the 2019 American Association of Physical Anthropologists statement on race and racism, http://physanth.org/about/position-statements/aapa-statement-race-and-racism-2019/.

   "ASHG Perspective: ASHG Denounces Attempts to Link Genetics and Racial Supremacy." *American Journal of Human Genetics* 103 (2018): 636. https://doi.org/10.1016/j.ajhg.2018.10.011.
10. Marks, "Ten Facts about Human Variation"; Gravlee, Non, and Mulligan, "Genetic Ancestry"; Fuentes, *Race*; Oluo, *So You Want to Talk about Race?*; Roberts, *Fatal Invention*; "ASHG Perspective"; http://www.understandingrace.com.
11. http://www.pewresearch.org/topics/discrimination-and-prejudice/, http://www.pewresearch.org/topics/income-inequality/.
12. http://www.pewresearch.org/topics/immigration-attitudes/2018/.
13. http://www.pewresearch.org/topics/income-inequality/, https://healthinequality.org/, Patten, "Racial, Gender Wage Gaps"; Oxfam International, "5 Shocking Facts."
14. Fuentes, *Creative Spark*.
15. Marks, *Why I Am Not a Scientist*.
16. See also Boudry and Pigliucci, *Science Unlimited?*

17. See, for example, Antón, Malhi, and Fuentes, "Race and Diversity in U.S. Biological Anthropology"; "Science Benefits from Diversity"; and the *Nature* and *Scientific American* joint issue on science and diversity: https://www.nature.com/news/diversity-1.15913.

18. Marks, *Why I Am Not a Scientist*; Marks, *Is Science Racist?*; Boudry and Pigliucci, *Science Unlimited?*

19. See, for example, Pinker, *Enlightenment Now*, as well as books, articles, blogs, and their accompanying assertions by Sam Harris, Richard Dawkins, Jerry Coyne, et al.

20. Dawkins, *God Delusion*; Coyne, *Faith vs. Fact.*

21. http://www.monkeyforestubud.com/concept-of-monkey-forest/the-tri-hita-karana/.

22. Fuentes, "Naturalcultural Encounters in Bali."

23. Snow, *Two Cultures.*

Bibliography

Abraham, A. "The Imaginative Mind." *Human Brain Mapping* 37 (2016): 4197–211.

Abraham, A. "Introduction." In *Cambridge Handbook of the Imagination,* ed. A. Abraham. Cambridge: Cambridge University Press [in press].

Ackermann, R. R., A. Mackay, and M. L. Arnold. 2015. "The Hybrid Origin of "Modern" Humans." *Journal of Evolutionary Biology* 43 (2015): 1–11. doi:10.1007/s11692-015-9348-1.

Adovasio, J. M., O. Soffer, and J. Page. *The Invisible Sex: Uncovering the True Roles of Women in Prehistory.* Washington, DC: Smithsonian Books, 2007.

Aiello, L. C., and S. C. Antón. 2012. "Human Biology and the Origins of *Homo*: An Introduction to Supplement 6." *Current Anthropology* 53, no. 56 (2012): S269–S277.

Alós-Ferrer, C. "A Review Essay on Social Neuroscience: Can Research on the Social Brain and Economics Inform Each Other?" *Journal of Economic Literature* 56, no. 1 (2018): 234–64. https://doi.org/10.1257/jel.20171370.

American Association of Physical Anthropologists. "AAPA Statement on Race & Racism." March 8, 2019, physanth.org/about/position-statements/aapa-statement-race-and-racism-2019.

American Economic Association. "What Is Economics?" https://www.aeaweb.org/resources/students/what-is-economics.

Andersson, Claes, A. Törnberg, and P. Törnberg. "An Evolutionary Developmental Approach to Cultural Evolution." *Current Anthropology* 55, no. 2 (2014): 154–74.

Anton, S. C., R. S. Malhi, and A. Fuentes. "Race and Diversity in U.S. Biological Anthropology: A Decade of AAPA Initiatives." *Yearbook of Physical Anthropology* 165, S65 (2018): 158–80. DOI: 10.1002/ajpa.23382.

Antón, S. C., R. Potts, and L. C. Aiello. "Evolution of Early *Homo*: An Integrated Biological Perspective." *Science* 345 (2014). doi:10.1126/science.1236828.

Antón, S. C., and J. J. Snodgrass. "Origin and Evolution of Genus *Homo*: New Perspectives." *Current Anthropology* 53, no. S6 (2012): S479–S496.

Arbib, M. A. "From Mirror Neurons to Complex Imitation in the Evolution of Language and Tool Use." *Annual Review of Anthropology* 40 (2011): 257–73.

Archer, W., D. R. Braun, J. W. K. Harris, J. T. McCoy, and B. G. Richmond. "Early Pleistocene Aquatic Resource Use in the Turkana Basin." *Journal of Human Evolution* 77 (2014): 74–87.

"ASHG Perspective: ASHG Denounces Attempts to Link Genetics and Racial Supremacy." *American Journal of Human Genetics* 103 (2018): 636. https://doi.org/10.1016/j.ajhg.2018.10.011

Atran, S. *In Gods We Trust: The Evolutionary Landscape of Religion.* Oxford: Oxford University Press, 2002.

Banning, E. B. 2015. "So Fair a House: Göbekli Tepe and the Identification of Temples in the Pre-Pottery Neolithic of the Near East." *Current Anthropology* 52, no. 5 (October 2011): 619–60.

Barash, D. P., and J. E. Lipton. *The Myth of Monogamy.* New York: Holt Paperbacks, 2001.

Barnard, A. *Social Anthropology and Human Origins.* Cambridge: Cambridge University Press, 2011.

Barker, J. "Happy Birthday, Karl Marx. You Were Right!" The New York Times, April 30, 2018, https://www.nytimes.com/2018/04/30/opinion/karl-marx-at-200-influence.html.

Barrett, J. L. *Why Would Anyone Believe in God?* Walnut Creek, CA: AltaMira Press, 2004.

Barrett, L. "Why Brains Are Not Computers, Why Behaviorism Is Not Satanism, and Why Dolphins Are Not Aquatic Apes." *Journal of Applied Behavioral Analysis* 39, no. 1 (November 2015): DOI 10.1007/s40614-015-0047-0.

Barton, R. A., and C. Venditti. "Rapid Evolution of the Cerebellum in Humans and Other Great Apes." *Current Biology* 24 (2014): 2440–44.

Bar-Yosef, O. "The Natufian Culture in the Levant: Threshold to the Origins of Agriculture." *Evolutionary Anthropology* (1998): 159–77.

Bateson, P., N. Cartwright, J. Dupré, K. Laland, and D. Noble. "New Trends in Evolutionary Biology: Biological, Philosophical, and Social Science Perspectives." *Interface Focus* 7 (2017): 20170051. http://dx.doi.org/10.1098/rsfs.2017.0051.

Bekoff, M. "Make No Mistake, Orca Mom J-35 and Pod Mates are Grieving." Psychology Today, August 1, 2018, https://www.psychologytoday.com/us/blog/animal-emotions/201808/make-no-mistake-orca-mom-j-35-and-pod-mates-are-grieving.

Bellah, R. N. *Religion in Human Evolution: From the Paleolithic to the Axial Age.* Cambridge, MA: Harvard University Press, 2011.

Bentley, G. R., T. Goldberg, and G. Jasienska. "The Fertility of Agricultural and Non-Agricultural Traditional Societies." *Population Studies* 47 (1993): 269–81.

Bering, J. *The Belief Instinct.* New York: W. W. Norton, 2011.

Bloch, M. "Why Religion Is Nothing Special but Is Central." *Philosophical Transactions of the Royal Society B: Biological Sciences* 363, no. 1499 (2008): 2055–61.

Blaug, M. "Economics." Encyclopaedia Britannica. Accessed May 7, 2019. https://www.britannica.com/topic/economics#ref236757.

Boehm, C. *Hierarchy in the Forest: The Evolution of Egalitarian Behavior.* Cambridge, MA: Harvard University Press, 1999.

Boehm, C. "What Makes Humans Economically Distinctive? A Three-Species Evolutionary Comparison and Historical Analysis." *Journal of Bioeconomics* 6 (2004): 109–35.

Bolduc, J., and F. Cezilly. "Optimality Modelling in the Real World." *Biology and Philosophy* 27 (2012): 851–69. DOI 10.1007/s10539-012-9333-3.

Bolnick, D. A. "Individual Ancestry Inference and the Reification of Race as a Biological Phenomenon." In *Revisiting Race in a Genomic Age*, edited by B. Koenig, S. Lee, and S. Richardson, 70–88. New Brunswick, NJ: Rutgers University Press.

Borgerhoff Mulder, M., S. Bowles, T. Hertz, et al. "Intergenerational Wealth Transmission and the Dynamics of Inequality in Smallscale Societies." *Science* 326 (2009): 682–88.

Borgerhoff Mulder, M., I. Fazzio, W. Irons, et al. "Pastoralism and Wealth Inequality: Revisiting an Old Question." *Current Anthropology* 51 (2010): 35–48.

Boudry, M., and M. Pigliucci, eds. *Science Unlimited? The Challenges of Scientism.* Chicago: University of Chicago Press, 2017.

Bourdieu, P. *Outline of a Theory of Practice.* Cambridge: Cambridge University Press, 1977.

Bowlby, J. *Attachment. Attachment and Loss.* Volume 1. *Loss.* New York: Basic Books, 1969.

Bowles, S., and J. K. Choi. "Coevolution of Farming and Private Property during the Early Holocene." *PNAS* 110, no. 22 (2013): 8830–35.

Bowles, S., and H. Gintis. *A Cooperative Species: Human Reciprocity and Its Evolution.* Princeton, NJ: Princeton University Press, 2011.

Boyd, R., P. J. Richerson, and J. Henrich. "The Cultural Niche: Why Social Learning Is Essential for Human Adaptation." *PNAS* 108, no. 52 (2011): 10918–25.

Boyer, P. *The Naturalness of Religious Ideas: A Cognitive Theory of Religion.* Berkeley, CA: University of California Press, 1994.

Boyer, P., and B. Bergstrom. "Evolutionary Perspectives on Religion." *Annual Review of Anthropology* 37 (2008): 111–30. doi: 10.1146/annurev.anthro.37 .081407.085201.

Boyer, P., and M. B. Peterson. "Folk-Economic Beliefs: An Evolutionary Cognitive Model." *Behavioral and Brain Sciences* 41 (2018): doi:10.1017/ S0140525X17001960, e158.

Brooks, A. S., et al. "Long-Distance Stone Transport and Pigment Use in the Earliest Middle Stone Age." *Science* 360, no. 6384 (2018): 10.1126/science .aao2646.

Brumann, C. "Writing for Culture: Why a Successful Concept Should Not Be Discarded." *Current Anthropology* 40 (1999): S1–S27.

Burkart, J. M., S. B. Hrdy, and C. van Schaik. "Cooperative Breeding and Human Cognitive Evolution." *Evolutionary Anthropology* 18 (2009): 175–86.

Callaway, E. "Domestication: The Birth of Rice." *Nature* 514 (2014): S58–S59.

Campbell, C., A. Fuentes, K. C. MacKinnon, S. Bearder, and R. Stumpf. *Primates in Perspective.* 2nd edition. New York: Oxford University Press, 2011.

Carbonell, E., and M. Mosquera. "The Emergence of a Symbolic Behaviour: The Sepulchral Pit of Sima de los Huesos, Sierra de Atapuerca, Burgos, Spain." *Comptes Rendus Palévol* 5 (2006): 155–60.

Carlén, M. "What Constitutes the Prefrontal Cortex?" *Science* 358, no. 6362 (2017): 478–82.

Carrier, J., and D. Miller. *Virtualism: A New Political Economy.* London: Bloomsbury, 1999.

Carter, C. S., and E. B. Keverne. "The Neurobiology of Social Affiliation and Pair Bonding." In *Hormones, Brain and Behavior*, 2nd ed., edited by D. W. Pfaff et al., 137–65. San Diego: Elsevier, 2009.

Chapais, B. *Primeval Kinship: How Pair-Bonding Gave Birth to Human Society.* Cambridge, MA: Harvard University Press, 2008.

Clare, L., et al. "Establishing Identities in the Proto-Neolithic: 'History-Making' at Göbekli Tepe from the Late Tenth Millennia cal BCE." In *Religion, History, and Place in the Origins of Settled Life*, edited by I. Hodder. Louisville: University Press of Colorado, 2018.

Clarkin, P. "Developmental Plasticity and the 'Hard-Wired' Problem. Blog. June 10, 2014. https://kevishere.com/2014/06/10/developmental-plasti city-and-the-hard-wired-problem.

Coyne, J. *Faith vs. Fact: Why Science and Religion Are Incompatible.* New York: Viking Penguin, 2015.

Crane, T. *The Meaning of Belief: Religion from an Atheist's Point of View.* Cambridge, MA: Harvard University Press, 2017.

Curtis, J. T., and Z. Wang. "The Neurochemistry of Pair Bonding." *Current Directions in Psychological Science* 12, no. 2 (2003): 49–53.

Davies, N., J. Krebs, and S. West. *Introduction to Behavioral Ecology.* 4th edition. Hoboken, NJ: John Wiley & Sons, 2012.

Davis, S. J. M., and F. Valla. "Evidence for Domestication of the Dog 12,000 Years Ago in the Natufian of Israel." *Nature* 276 (1978): 608–10.

Dawkins, R. *The God Delusion.* New York: Bantam Books, 2006.

Deacon, T. "On Human (Symbolic) Nature: How the Word Became Flesh." In

Embodiment in Evolution and Culture, edited by T. Fuchs and C. Tewes, 129–49. Tübingen, Germany: Mohr Siebeck, 2016.

Deino, A. L., et al. "Chronology of the Acheulean to Middle Stone Age Transition in Eastern Africa." *Science* 360, no. 6384 (April 6, 2018): 95–98. 10.1126/science.aao2216.

Dirks, P. H. G. M., L. R. Berger, E. M. Roberts, J. D. Kramers, J. Hawks, P. S. Randolph-Quinney, M. E. Elliott, et al. "Geological and Taphonomic Context for the New Hominin Species Homo naledi from the Dinaledi Chamber, South Africa." Elife (2015). https://doi.org/10.7554/eLife.09561.001.

Dirks, P. H., E. M. Roberts, H. Hilbert-Wolf, et al. "The Age of *Homo naledi* and Associated Sediments in the Rising Star Cave, South Africa." *eLife* (2017): doi:10.7554/eLife.24231.

Donald, M. *A Mind So Rare: The Evolution of Human Consciousness.* New York: W. W. Norton, 2001.

Donald, M. *The Origins of the Modern Mind.* Cambridge, MA: Harvard University Press, 1991.

Downey, G. and D. H. Lende. "Evolution and the Brain." In *The Encultured Brain*, edited by D. H. Lende and Greg Downey, 103–38. Cambridge, MA: MIT Press, 2012.

Dugatkin, L., and L. Trut. *How to Tame a Fox (and Build a Dog): Visionary Scientists and a Siberian Tale of Jump-Started Evolution.* Chicago: University of Chicago Press, 2017.

Durkheim, E. *The Elementary Forms of the Religious Life.* Translated by Karen Fields. New York: Free Press, 2005 (1912).

Durkheim, E. *The Elementary Forms of Religious Life.* Oxford: Oxford University Press, 2008 (1912).

Eliade, M. *The Sacred and the Profane: The Nature of Religion.* Translated from the French by W. R. Trask. New York: Harcourt, 1987 (1957).

Ellison, P. T., and P. B. Gray, eds. *The Endocrinology of Social Relationships.* Cambridge, MA: Harvard University Press, 2009.

Estalrrich, A., and A. Rosas. "Division of Labor by Sex and Age in Neandertals: An Approach through the Study of Activity-Related Dental Wear." *Journal of Human Evolution* 80 (2015): 51–63.

Estrada, A. et al. (30 authors, incl. A. Fuentes). "Impending Extinction Crisis of the World's Primates: Why Primates Matter." *Science Advances* 3, no. 1 (2017): e1600946, doi: 10.1126/sciadv.1600946.

Fausto-Sterling, A. *Sex/Gender: Biology in a Social World.* New York: Routledge, 2012.

Fausto-Sterling, A. *Sexing the Body: Gender Politics and the Construction of Sexuality.* New York: Basic Books, 2000.

Ferguson, B. R. "Born to Live: Challenging Killer Myths." In *Origins of Altruism and Cooperation*, edited by R. W. Sussman and C. R. Cloninger, 249–70 (Developments in Primatology Series 36, Part 2: Progress and Prospects). New York: Springer, 2011.

Ferguson, B. R. "Pinkers List: Exaggerating Prehistoric War Mortality." In *War, Peace and Human Nature*, edited by D. Fry, 112–31. New York: Oxford University Press, 2013.

Ferguson, B. R. "War before History." In *The Ancient World at War*, edited by P. D'Souza, 15–27. London: Thames and Hudson, 2008.

Fine, C. *Delusions of Gender: How Our Minds, Society, and Neurosexism Create Difference*. W. W. Norton, 2011.

Fisher, H. *Why We Love: The Nature and the Chemistry of Romantic Love*. New York: Owl Books, 2004.

Flannery, K. V. "The Origins of Agriculture." *Annual Review of Anthropology* 2 (1973): 271–310.

Fleischacker, S. "Adam Smith's Moral and Political Philosphy." In The Stanford Encyclopedia of Philosophy, edited by E. N. Zalta, Spring 2017 ed. https ://plato.stanford.edu/entries/smith-moral-political.

Fletcher, G. J. O., J. A. Simpson, L. Campbell, and N. C. Overall. "Pair-Bonding, Romantic Love, and Evolution: The Curious Case of *Homo Sapiens*." *Perspectives on Psychological Science* 10, no. 1 (2015): 20–36.

Foley, R. A. "Mosaic Evolution and the Pattern of Transitions in the Hominin Lineage." *Philosophical Transactions of the Royal Society B: Biological Sciences* 371 (2016): 20150244. doi:10.1098/rstb.2015.0244.

Foote, A. D., et al. "Genome-Culture Coevolution Promote Rapid Divergence of Killer Whale Ecotypes." *Nature Communications* 7 (2016): 11693. DOI:10.1038/ncomms11693.

Fox-Keller, E. *The Century of the Gene*. Cambridge, MA: Harvard University Press, 2002.

Fry, D. P. *The Human Potential for Peace: An Anthropological Challenge to Assumptions about War and Violence*. New York: Oxford University Press, 2006.

Fry, D. P., ed. *War, Peace, and Human Nature*. New York: Oxford University Press, 2013.

Fuentes, A. *Biological Anthropology: Concepts and Connections*. 3rd edition. New York: McGraw-Hill, 2018.

Fuentes, A. *The Creative Spark: How Imagination Made Humans Exceptional*. New York: Dutton/Penguin, 2017.

Fuentes, A. *Evolution of Human Behavior*. New York: Oxford University Press, 2009.

Fuentes, A. "The Extended Evolutionary Synthesis, Ethnography, and the Human Niche: Toward an Integrated Anthropology." *Current Anthropology* 57, Supplement 13 (2016): 13–26. DOI: 10.1086/685684.

Fuentes, A. "How Humans and Apes Are Different, and Why It Matters." *Journal of Anthropological Research* 74, no. 2 (2018): 151–67. doi.org/10.1086/697150.

Fuentes, A. "Human Evolution, Niche Complexity, and the Emergence of a Distinctively Human Imagination." *Time and Mind* 7, no. 3 (2014): 241–57. DOI :10.1080/1751696X.2014.945720.

Fuentes, A. "Human Niche, Human Behaviour, Human Nature." *Interface Focus* 7 (2017): 20160136. http://dx.doi.org/10.1098/rsfs.2016.0136.

Fuentes, A. "Hylobatid Communities: Changing Views on Pair Bonding and Social Organization in Hominoids." *Yearbook of Physical Anthropology* 113, no. s31(2000): 33–60.

Fuentes, A. "Integrative Anthropology and the Human Niche: Toward a Contemporary Approach to Human Evolution." *American Anthropologist* 117, no. 2(2015): 302–15. DOI: 10.1111/aman.12248.

Fuentes, A. "Naturalcultural Encounters in Bali: Monkeys, Temples, Tourists, and Ethnoprimatology." *Cultural Anthropology* 25, no. 4 (2010): 600–624.

Fuentes, A. "Patterns and Trends in Primate Pair Bonds." *International Journal of Primatology* 23, no. 5 (2002): 953–78.

Fuentes, A. *Race, Monogamy, and Other Lies They Told You: Busting Myths about Human Nature.* Berkeley: University of California Press, 2012.

Fuentes, A. "Re-Evaluating Primate Monogamy." *American Anthropologist* 100, no. 4 (1999): 890–907.

Fuentes, A. "Social Systems and Socioecology: Understanding the Evolution of Primate Behavior." In *Primates in Perspective*, 2nd edition, edited by C. Campbell, A. Fuentes, K. MacKinnon, S. Bearder, and R. Stumpf, 500–511. New York: Oxford University Press, 2001.

Fuentes, A., and P. Wiessner. "Reintegrating Anthropology: From Inside Out." *Current Anthropology* 57, Supplement no. s13 (2016): 3–12.

Fuentes, A., M. Wyczalkowski, and K. C. MacKinnon. "Niche Construction through Cooperation: A Nonlinear Dynamics Contribution to Modeling Facets of the Evolutionary History in the Genus *Homo*." *Current Anthropology* 51, no. 3 (2010): 435–44.

Fuller, D. Q., et al. "Convergent Evolution and Parallelism in Plant Domestication Revealed by an Expanding Archaeological Record." *PNAS* 111, no. 17 (2014): 6147–52.

Fuller, D. Q., et al. "The Domestication Process and Domestication Rate in Rice: Spikelet Bases from the Lower Yangtze." *Science* 323 (2019): 1607–10.

Fuller, D. Q., and A. Weisskopf. "The Early Rice Project: From Domestication to Global Warming." *Archaeology International* no. 13/14 (2009–2011): 44–51. DOI: http://dx.doi.org/10.5334/ai.1314.

Galor, O., and O. Moav. "Natural Selection and the Origin of Economic

Growth." *Quarterly Journal of Economics* (Oxford University Press) 117, no. 4 (2002): 1133–91.

Gamble, C., J. Gowlett, and R. Dunbar. "The Social Brain and the Shape of the Palaeolithic." *Cambridge Archaeological Journal* 21, no. 1 (2011): 115–36.

Garderinck, I., and M. Osvath. "The Tripod Effect: Co-Evolution of Cooperation, Cognition, and Communication." In *The Symbolic Species Evolved*, edited by T. Schilhab, F. Stjernfelt, and T. Deacon, 193–224. New York: Springer, 2012.

Geertz, C. "Religion as a Cultural System." In *Anthropological Approaches to the Study of Religion*, edited by M. Banton, 1–46. London: Tavistock, 1966.

Geertz, C. *Thick Description: Towards an Interpretive Theory of Culture*. New York: Basic Books, 1973.

Gero, J. M.,. and M. W. Conkey. *Engendering Archeology: Women and Prehistory*. Oxford: Blackwell, 1991.

Gettler, L. T. "Applying Socioendocrinology to Evolutionary Models: Fatherhood and Physiology." *Evolutionary Anthropology* 23, no. 4 (2014): 146–60.

Gettler, L. T. "Becoming DADS: Considering the Role of Cultural Context and Developmental Plasticity for Paternal Socioendocrinology." *Current Anthropology* 57 (2016): S38–S51.

Gilbert, S. F., J. Sapp, and A. I. Tauber. "A Symbiotic View of Life: We Have Never Been Individuals." *Quarterly Review of Biology* 87, no. 4 (2012): 325–41.

Goldman, J. G. "Is Language Unique to Humans?" BBC, October 17, 2012. http://www.bbc.com/future/story/20121016-is-language-unique-to-humans.

Goldschmidt, W. *The Bridge to Humanity: How Affect Hunger Trumps the Selfish Gene*. Oxford: Oxford University Press, 2005.

Gómez, J., M. Verdú, A. González-Megías, and M. Méndez-Gomez. "The Phylogenetic Roots of Human Lethal Violence." *Nature* 538 (2016): 233–37. doi:10.1038/nature19758.

Gots, J. "Humans Make Language, Language Makes Us Human." Big Think, September 22, 2011, https://bigthink.com/floating-university/humans-make-language-language-makes-us-human.

Gracia, A. et al. "Craniosynostosis in the Middle Pleistocene human Cranium 14 from the Sima de los Huesos, Atapuerca, Spain." *PNAS* 106, no. 16 (2009): 6573–78. DOI: 10.1073/pnas.0900965106

Graeber, D. "Beads and Money: Notes toward a Theory of Wealth and Power." *American Ethnologist* 23 (1996): 4–24.

Graeber, D. *Debt: The First 5000 Years*. Brooklyn, NY: Melville House, 2011.

Graeber, D. *Toward an Anthropological Theory of Value: The False Coin of Our Own Dreams*. New York: Palgrave, 2001.

Gravlee, C. C., A. L. Non, and C. J. Mulligan. "Genetic Ancestry, Social Classification, and Racial Inequalities in Blood Pressure in Southeastern Puerto Rico." *PloS ONE* 4(9) (2009): e6821. doi:10.1371/journal.pone.0006821.

Gray, P. B., and B. C. Campbell. "Human Male Testosterone, Pair-Bonding, and Fatherhood." In *Endocrinology of social relationships*, edited by P. B. Gray and P. T. Ellison, 270–93. Cambridge, MA: Harvard University Press, 2009.

Grove, M., and F. Coward. "Beyond the Tools: Social Innovation and Hominin Evolution." *Paleoanthropology* (January 2011): 111-29. doi:10.4207/PA.2011. ART46.

Grove, M., and F. Coward. "From Individual Neurons to Social Brains." *Cambridge Archeological Journal* 18 (2008): 387–400.

Gvozdover, M. D. "The Typology of Female Figurines of the Kostenki Paleolithic Culture." *Soviet Anthropology and Archaeology* 27 (1989): 32–94.

Gyatso, T. "Compassion and the Individual." Accessed May 7, 2019. https://www.dalailama.com/messages/compassion-and-human-values/compassion.

Ha, R. R. "Cost-Benefit Analysis." In *Encyclopedia of Animal Behavior*, edited by M. D. Breed and J. Moore, 1:402–5. Oxford: Academic Press, 2010.

Haas, J., and M. Piscitelli. "The Prehistory of Warfare: Misled by Ethnography." In *War, Peace, and Human Nature*, edited by D. P. Fry, 168–90. Oxford: Oxford University Press, 2013.

Hamilton, C., and J. Grinevald. "Was the Anthropocene Anticipated?" *Anthropocene Review* 2, no. 1 (2015): 59–72.

Hammer, M. F., A. E. Woerner, F. L. Mendez, J. C. Watkins, and J. D. Wall. "Genetic Evidence for Archaic Admixture in Africa." *PNAS* 108 (September 2011): 15123–28.

Han, S. *The Sociocultural Brain*. Oxford: Oxford University Press, 2017.

Han, S, and Y. Ma. "A Culture-Behavior-Brain Loop Model of Human Development." *Trends in Cognitive Sciences* 19 (2015): 666–76.

Haraway, D. "A Cyborg Manifesto: Science, Technology, and Socialist-Feminism in the Late Twentieth Century" In, *Simians, Cyborgs and Women: The Reinvention of Nature*, edited by D. Haraway, 149–81. New York: Routledge, 1991.

Haraway, D. *Simians, Cyborgs, and Women: The Reinvention of Nature*. New York: Routledge, 1991.

Hare, B. "From Hominoid to Hominid Mind: What Changed and Why?" *Annual Review of Anthropology* 40 (2011): 293–309.

Harmand. S., et al. "3.3-Million-Year-Old Stone Tools from Lomekwi 3, West Turkana, Kenya." *Nature* 521 (2015): 310–16.

Hart, D., and R. W. Sussman. *Man the Hunted: Primates, Predators, and Human Evolution*. Cambridge, MA: Westview Press, 2005.

Hart, K. "Notes towards an Anthropology of Money." *Kritikos* 2 (June 2005). https://intertheory.org/hart.htm.

Hatfield, E., and R. L. Rapson. "Love and Sex: Cross-Cultural Perspectives."
Lanham, MD: University Press of America, 2005.

Hausman, D. M. "Philosophy of Economics." In The Stanford Encyclopedia of
Philosophy, edited by E. N. Zalta, Fall 2018 ed. https://plato.stanford.edu/
entries/economics/#1.

Hawk, J. "What Is the 'Braided Stream' Analogy for Human Evolution?"
Weblog. November 26, 2015. http://johnhawks.net/weblog/topics/news/fin-
layson-braided-stream-2013.html.

Helm, B. "Love." In The Stanford Encyclopedia of Philosophy, edited by E. N.
Zalta, Fall 2017 ed. https://plato.stanford.edu/entries/love.

Helm, B. W. "Love, Identification, and the Emotions." American Philosophical
Quarterly 46 (2009): 39–59.

Henrich, J., R. Boyd, S. Bowles, C. Camerer, E. Fehr, and H. Gintis. Foundations of
Human Sociality: Economic Experiments and Ethnographic Evidence from Fifteen
Small-Scale Societies. Oxford: Oxford University Press, 2004.

Hiscock, P. "Learning in Lithic Landscapes: A Reconsideration of the Hominid
'Toolmaking' Niche." Biological Theory 9 (2014): 27–41.

Hobaiter, C., T. Poisot, K. Zuberbühler, W. Hoppit, and T. Gruber. 2014. "Social
Network Analysis Shows Direct Evidence for Social Learning of Tool Use in
Wild Chimpanzees." PLOS Biology 12 (2014): e1001960.

Hodder, I., ed. Religion, History, and Place in the Origin of Settled Life. Boulder:
University of Colorado Press, 2018. Hrdy, S. B. Mothers and Others: The Evolutionary Origins of Mutual Understanding. Cambridge, MA: Harvard University Press, 2009.

Hovers, E. "Tools to Go Back in Time." Nature 521 (May 2015): 294–95. https://
doi.org/10.1038/521294a.

Huang, X., et al. "A Map of Rice Genome Variation Reveals the Origin of Culti-
vated Rice." Nature 490 (October 25, 2012): 497–501. doi:10.1038/nature11532.

Hublin, J. J., A. Ben-Neer, S. E. Bailey, S. Freidline, M. S. Neubauer, M. M.
Skinner, et al. 2017. "New Fossils from Jebel Irhoud, Morocco, and the Pan-
African Origin of Homo Sapiens." Nature 546, no. 7657 (2017): 289–92.

Hunley, K. L., G. S. Cabana, and J. C. Long. "The Apportionment of Human
Diversity Revisited." American Journal of Physical Anthropology 160 (2016):
561–69.

Hunt, C. O., and R. J. Rabett. "Holocene Landscape Intervention and Plant
Food Production Strategies in Island and Mainland Southeast Asia." Journal
of Archaeological Science 51 (2014): 22–33.

Hutchinson, G. E. "Concluding Remarks" Cold Spring Harbor Symposium on
Quantitative Biology 22 (1957): 415–27.

Ingold, T. "The Art of Translation in a Continuous World." In Beyond Boundaries:

Understanding, Translation, and Anthropological Discourse, edited by G. Palsson, 210–30. London: Berghahn, 1993.

Ingold, T. "Evolution in a Minor Key." In *Verbs, Bones, and Brains*, edited by A. Fuentes and A. Visala. Notre Dame, IN: University of Notre Dame Press, 2017.

Ingold, T. *The Perception of the Environment: Essays on Livelihood, Dwelling, and Skill.* Abingdon, UK: Routledge, 2000.

Ingold, T. "To Human Is a Verb." In *Verbs, Bones, and Brains*, edited by A. Fuentes and A. Visala. Notre Dame, IN: University of Notre Dame Press, 2017.

Johannson, D. "Lucy, Thirty Years Later: An Expanded View of *Australopithecus Afarensis.*" *Journal of Anthropological Research* 60, no. 4 (2004): 466–68.

Johnson, D. D. P., and J. M. Bering. "Hand of God, Mind of Man: Punishment and Cognition in the Evolution of Cooperation." *Evolutionary Psychology* 4 (2006): 219–33.

Jones, B. "Growning Up Black in America: Here's My Story of Everday Racism." The Guardian, June 6, 2018. https://www.theguardian.com/us-news/2018/jun/06/growing-up-black-in-america-racism-education.

Kaplan, H. "A Theory of Fertility and Parental Investment in Traditional and Modern Human Societies." *American Journal of Physical Anthropology* 101, no. s23 (1996): 91–135.

Karandashev, V. "A Cultural Perspective on Romantic Love." *Online Readings in Psychology and Culture* 5, no. 4 (2015). https://doi.org/10.9707/2307-0919.1135.

Katz, Y. "Noam Chomsky on Where Artificial Intelligence Went Wrong." The Atlantic, Novermber 1, 2012.https://www.theatlantic.com/technology/archive/2012/11/noam-chomsky-on-where-artificial-intelligence-went-wrong/261637.

Kierkegaard, S. *Works of Love.* New York: Harper Perennial Modern Classics, 2009 (1847).

King, B. *How Animals Grieve.* Chicago: University of Chicago Press, 2013.

King, B. *The Information Continuum: Evolution of Social Information Transfer in Monkeys, Apes, and Hominids.* Santa Fe: School of American Research Press, 1994.

Kissel, B., and A. Fuentes. "'Behavioral Modernity' as a Process, Not an Event, in the Human Niche." Time and Mind 11, no. 2 (2018): 163–83. doi:10.1080/1751696X.2018.1469230.

Kissel, M., and A. Fuentes. "A Database of Archaeological Evidence for Representational Behavior." *Evolutionary Anthropology* 26, no. 4 (2017): 1490150. DOI: 10.1002/evan.21525.

Kissel, M., and A. Fuentes. "Semiosis in the Pleistocene." *Cambridge Archaeological Journal* 27, no. 3 (2017): 397–412. doi:10.1017/S0959774317000014.

Kissel, M., and N. Kim. "The Emergence of Human Warfare: Current Perspectives." *American Journal of Physical Anthropology* 168, no. 1 (2018): DOI: 10.1002/ajpa.23751.

Kissel, M., and N. Kim. *Emergent Warfare in Our Evolutionary Past.* Oxford: Routledge, 2018.

Koestler, A. *The Ghost in the Machine.* New York: Penguin Group, 1990 (1967).

Kokko, H. 2017. "Give One Species the Task to Come Up with a Theory That Spans Them All: What Good Can Come Out of That?" *Proceedings of the Royal Society B: Biological Sciences* 284 (2017): 20171652. http://dx.doi.org/10.1098/rspb.2017.1652.

Konner, M. Believers: Faith in Human Nature. New York: W. W. Norton, in press.

Konner, M. *The Tangled Wing: Biological Constraints on the Human Spirit.* New York: Harper, 1983.

Kramer, K. L., and E. Otarola-Castillo. "When Mothers Need Others: The Impact of Hominin Life History Evolution on Cooperative Breeding." *Journal of Human Evolution* 84 (July 2015): 16–24. doi.org/10.1016/j.jhevol.2015.01.009.

Kroeber, A. L., and Clyde Kluckhohn. *Culture: A Critical Review of Concepts and Definitions.* New York: Vintage, 1952.

Kuhn, S. L., and M. C. Stiner. "What's a Mother to Do? A Hypothesis about the Division of Labor among Neanderthals and Modern Humans in Eurasia." *Current Anthropology* 47 (2006): 953–80.

Kuijt, I. "What Do We Really Know about Food Storage, Surplus, and Feasting in Preagricultural Communities?" *Current Anthropology* 50, no. (2009): 641–44.

Kuijt, I., and A. M. Prentiss. "Niche Construction, Macroevolution, and the Late Epipaleolithic of the Near East." In *Macroevolution in Human Prehistory,* edited by A. M. Prentiss, I. Kuijt, and J. Chatters, 253–71. New York: Springer, 2009.

Kuzawa, C., et al. "Metabolic Costs and Evolutionary Implications of Human Brain Development." *PNAS* 111, no. 36 (2014): 13010–15.

Kuzawa, C. W., and J. M. Bragg. "Plasticity in Human Life History Strategy: Implications for Contemporary Human Variation and the Evolution of Genus *Homo.*" *Current Anthropology* 53, S6 (2012): S369–S382. Special issue: "Human Biology and the Origins of *Homo.*"

Laland, K. N. *Darwin's Unfinished Symphony: How Culture Made the Human Mind.* Princeton, NJ: Princeton University Press, 2017.

Laland, K. N., T. Uller, M. Feldman, K. Sterelny, G. B. Müller, A. Moczek, E. Jabonka, and J. Odling-Smee. "Does Evolutionary Theory Need a Rethink? Yes, Urgently." *Nature* 514, no. 7521 (2014): 161–64.

Laland, K. N., T. Uller, M. W. Feldman, K. Sterelny, G. B. Müller, A. Moczek, E. Jablonka, and J. Odling-Smee. "The Extended Evolutionary Synthesis: Its Structure, Assumptions, and Predictions." *Proceedings of the Royal Society B: Biological Sciences* 282 (2015): 20151019. http://dx.doi.org/10.1098/rspb.2015.1019.

Larsen, C. S. "Biological Changes in Human Populations with Agriculture." *Annual Review of Anthropology* 24 (1995): 185–213.

Larson, G., and D. Fuller. "The Evolution of Animal Domestication." *Annual Review of Ecology, Evolution, and Systematics* 45 (2014): 115–36.

Lende, Daniel H., and Greg Downey, eds. *The Encultured Brain: An Introduction to Neuroanthropology.* Cambridge, MA: MIT Press, 2012.

Lewis, C. S. *The Four Loves.* New York: Harcourt, 1971.

Liszkowski, U. "Emergence of Shared Reference and Shared Minds in Infancy." *Current Opinion in Psychology* 23 (2018): 26–29.

Lordkipanidze, D., et al. "A Complete Skull from Dmanisi, Georgia, and the Evolutionary Biology of Early *Homo*." *Science* 342, no. 6156 (2013): 326–31. DOI: 10.1126/science.1238484.

Lovejoy, O. "Reexamining Human Origins in Light of *Ardipithecus ramidus*." *Science* 326 (2009): 108–15 .

Luhby, T. "71% of the World's Population Lives on Less Than $10 a Day." CNN Business, July 8, 2015, https://money.cnn.com/2015/07/08/news/economy/global-low-income/index.html.

Luhrmann, T. M. et al. "Toward an Anthropological Theory of Mind." *Suomen Antropologi: Journal of the Finnish Anthropological Society* 36, no. 4 (Winter 2011): 5–69.

Macintosh, A. A., R., Pinhasi, and J. T. Stock "Prehistoric Women's Manual Labor Exceeded That of Athletes through the First 5,500 Years of Farming in Central Europe." *Science Advances* 3, no. 11 (2017): eaao3893. DOI: 10.1126/sciadv.aao3893.

Macintyre, A. *Dependent Rational Animals: Why Human Beings Need the Virtues.* Peru, IL: Open Court Press, 2001.

Maher, L. A., J. T. Stock, S. Finney, J. J. N. Heywood, P. T. Miracle, et al. "A Unique Human-Fox Burial from a Pre-Natufian Cemetery in the Levant (Jordan)." *PLoS ONE* 6, no. 1 (2011): e15815. doi:10.1371/journal.pone.0015815.

Marean, C. W. "An Evolutionary Anthropological Perspective on Modern Human Origins." *Annual Review of Anthropology* 44 (2015): 533–56.

Marks, J. *Is Science Racist?* Cambridge: Polity Press, 2016.

Marks, J. "Ten Facts about Human Variation." In *Human Evolutionary Biology*, edited by M. Muehlenbein, 265–76. New York: Cambridge University Press, 2010.

Marks, J. *Why I Am Not a Scientist: Anthropology and Modern Knowledge*. Berkeley: University of California Press, 2009.

Martine, W. *Untrue*. New York: Little-Brown, 2018.

Martinez, I., M. Rosa, J.-L. Arsuaga, P. Jarabo, R. Quam, C. Lorenzo, A. Gracia, J.-M. Carretero, J.-M. B. de Castro, and E. Carbonell. "Auditory Capacities in Middle Pleistocene Humans from the Sierra de Atapuerca in Spain." *PNAS* 101, no. 27 (2004): 9976–81. doi:10.1073/pnas.0403595101.

Marx, K. *The Poverty of Philosophy*. Paris, 1847. https://www.marxists.org/archive/marx/works/1847/poverty-philosophy/

Masci, D., and D. DeSilver. "A Global Snapshot of Same-Sex Marriage." Pew Research Center. December 8, 2017. https://www.pewresearch.org/fact-tank/2017/12/08/global-snapshot-sex-marriage.

Mattison, S., E. A. Smith, M. K. Shenk, and E. Cochrane. "The Evolution of Inequality." *Evolutionary Anthropology* 25 (2016): 184–99.

Maurer, B. "The Anthropology of Money." *Annual Review of Anthropology* 35 (2006): 15–36.

Mauss, M. *The Gift: The Form and Reason for Exchange in Archaic Societies*. London: Routledge, 2002 (1954).

McBrearty, S., and A. Brooks. "The Revolution That Wasn't: A New Interpretation of the Origin of Modern Human Behavior." *Journal of Human Evolution* 39 (2000): 453–63.

McDermott, L. "Self-Representation in Upper Paleolithic Female Figurines." *Current Anthropology* 37, no. 2 (1996): 227–75.

McNamara, P. *The Neuroscience of Religious Experience*. Cambridge: Cambridge University Press, 2009.

McPherron, S. P., Z. Alemseged, C. W. Marean, J. G. Wynn, D. Reed, D. Geraads, R. Bobe, and H. A. Béarat. "Evidence for Stone-Tool-Assisted Consumption of Animal Tissues before 3.39 Million Years Ago at Dikika, Ethiopia." *Nature* 466 (2010): 857–60.

Montagu, A. *The Human Revolution*. New York: Bantam, 1965.

Morgan, T. J. H., N. T. Uomini, L. E. Rendell, L. Chouinard-Thuly, S. E. Street, H. M. Lewis, C. P. Cross, et al. "Experimental Evidence for the Co-Evolution of Hominin Tool-Making Teaching and Language." *Nature Communications* 6 (January 13, 2015): 6029. DOI: 10.1038/ncomms7029.

Noë, R., and P. Hammerstein. "Biological Markets." *TREE* 10, no. 8 (1995): 336–39.

Norenzayan, A. *Big Gods: How Religion Transformed Cooperation and Conflict*. Princeton, NJ: Princeton University Press, 2013.

Norenzayan, A. "Does Religion Make People Moral?" *Behaviour* 151 (2014): 365–84.

Norenzayan, A., J. Henrich, and E. Slingerland. "Religious Prosociality: A Synthesis." In *Cultural evolution*, edited by P. Richerson and M. Christiansen, 365–78. Cambridge, MA: MIT Press, 2013.

Nussbaum, M. "Love and the Individual: Romantic Rightness and Platonic Aspiration." In M. Nussbaum, *Love's Knowledge: Essays on Philosophy and Literature*, 314–34. Oxford: Oxford University Press, 1990.

O'Brien, M. J., and K. N. Laland. "Genes, Culture, and Agriculture: An Example of Human Niche Construction." *Current Anthropology* 53, no. 4 (2012): 434–70.

Oka, R., and A. Fuentes. "From Reciprocity to Trade: How Cooperative Infrastructures Form the Basis of Human Socioeconomic Evolution." In *Cooperation in Social and Economic Life*, edited by R. C. Marshall, 3–28. Lanham, MD: Altamira Press, 2010.

Olmert, M. D. *Made for Each Other: The Biology of the Human-Animal Bond*. Philadelphia: Da Capo Press, 2009.

Oluo, I. *So You Want to Talk about Race?* New York: Seal Press, 2018.

Oxfam International. "5 Shocking Facts About Extreme Global Inequality and How to Even It Up." Accessed May 8, 2109. https://www.oxfam.org/en/even-it/5-shocking-facts-about-extreme-global-inequality-and-how-even-it-davos.

Parker, G., and J. Smith. "Optimality Theory in Evolutionary Biology." *Nature* 348, no. 6296 (November 1, 1990): 27–33.

Patten, E. "Racial, Gender Wage Gaps Persist in U.S. Despite Some Progress." Pew Research Center. July, 1, 2016. http://www.pewresearch.org/fact-tank/2016/07/01/racial-gender-wage-gaps-persist-in-u-s-despite-some-progress.

Peirce, C. S. *Collected Papers of Charles Sanders Peirce*. Volume 8. Edited by A. W. Burks. Cambridge, MA: Harvard University Press, 1958.

Peirce, C. S. *The Essential Peirce*. Volume 2. Bloomington: Indiana University Press, 1998.

Peirce, C. S. *The Logic of Interdisciplinarity*. 1st edition. The Monist Series (Deutsche Zeitschrift Fur Philosophie. Sonderband). n.p.: Akademie Verlag, 2009.

Perelman, P., W. E. Johnson, C. Roos, H. N. Seuánez, J. E. Horvath, M. A. M. Moreira, et al. 2011. "A Molecular Phylogeny of Living Primates." *PLoS Genetics* 7, no. 3 (2011): e1001342. https://doi.org/10.1371/journal.pgen.1001342.

Pinker, S. *The Better Angels of Our Nature: Why Violence Has Declined*. New York: Penguin, 2012.

Pew Research Center Forum on Religion and Public Life. "Executive Summary." In The Global Religious Landscape, edited by T. Miller. Washington, DC: Pew Research Center, 2012.

Pinker, S. *Enlightenment Now.* New York: Viking, 2018.

Pope, M., K. Russel, and K. Watson. "Biface Form and Structured Behavior in the Acheulean." *Lithics* 27 (2006): 44–57.

Potts, R. "Environmental and Behavioral Evidence Pertaining to the Evolution of Early *Homo*." *Current Anthropology* 53, no. S6 (December 2012): S299–S317.

Quinlan, R. "Human Pair-Bonds: Evolutionary Functions, Ecological Variation, and Adaptive Development." *Evolutionary Anthropology* 17 (2008): 227–38.

Ramsey, G. "Culture in Humans and Other Animals." *Biol Philos* 28 (2013): 457–79. DOI 10.1007/s10539-012-9347-x.

Rappaport, R. A. *Ritual and Religion in the Making of Humanity.* Cambridge: Cambridge University Press, 1999.

Read, D. *How Culture Makes Us Human: Primate Social Evolution and the Formation of Human Societies.* Walnut Creek, CA: Left Coast, 2011.

Rice, P. C. "Prehistoric Venuses: Symbols of Motherhood or Womanhood?" *Journal of Anthropological Research* 37 (1981): 402–14.

Richerson, P., et al. "Cultural Group Selection Plays an Essential Role in Explaining Human Cooperation: A Sketch of the Evidence." *Behavioral and Brain Sciences* 39 (2016): 55. doi:10.1017/S0140525X1400106X.

Richerson, P., and R. Boyd. *Not by Genes Alone: How Culture Transformed Human Evolution.* Chicago: University of Chicago Press, 2005.

Rilling, J. K. "The Neural and Hormonal Bases of Human Parental Care." *Neuropsychologia* 51, no. 4 (2013): 731–47.

Ripple, W. J., C. Wolf, T. M. Newsome, M. Galetti, et al. "World Scientists' Warning to Humanity: A Second Notice." *BioScience* 67, no. 12 (December 1, 2017): 1026–28, https://doi.org/10.1093/biosci/bix125.

Roberts, D. *Fatal Invention: How Science, Politics, and Big Business Re-Create Race in the Twenty-First Century.* New York: New Press, 2011.

Rodseth, L. "Hegemonic Concepts of Culture: The Checkered History of Dark Anthropology." *American Anthropologist* 120, no. 3 (2018): DOI: 10.1111/aman.13057.

Rossano, M. J. "Ritual Behavior and the Origin of Modern Cognition." *Cambridge Archaeological Journal* 19, no. 2 (2009): 243–56.

Ryan, C., and C. Jetha. *Sex at Dawn: The Prehistoric Origins of Modern Sexuality.* New York: Harper Books, 2010.

Sala, N., et al. "Lethal Interpersonal Violence in the Middle Pleistocene." *Plos ONE* 10 (2015): e0126589. DOI: 10.1371/journal.pone.0126589.

Sanz, C. M., J. Call, and C. Boesch. *Tool Use in Animals: Cognition and Ecology.* Cambridge: Cambridge University Press, 2014.

Scerri, E. et al. "Did Our Species Evolve in Subdivided Populations across Africa, and Why Does It Matter?" *Trends in Geology & Evolution* 33, no. 8 (2018): 582–94. https://doi.org/10.1016/j.tree.2018.05.005.

"Science Benefits from Diversity" (editorial). *Nature* 558, no. 5 (2018): doi: 10.1038/d41586-018-05326-3.

Shea, J. J.. "*Homo Sapiens* Is as *Homo Sapiens* Was." *Current Anthropology* 52, no. 1 (2011): 1–35.

Sheehan, O., J. Watts, R. D. Gray, and Q. D. Atkinson. "Coevolution of Landesque Capital-Intensive Agriculture and Sociopolitical Hierarchy." *PNAS* 115, no. 14 (2018): 3628n33. doi.org/10.1073/pnas.1714558115.

Sherrington, C. S. "Man on His Nature." The Gifford Lectures. Accessed May 8, 2019. https://www.giffordlectures.org/lectures/man-his-nature.

Sherwood, C. C., and A. Gomez-Robles. "Brain Plasticity and Human Evolution." *Annual Review of Anthropology* 46 (2017): 399–419.

Shipman, P. *The Animal Connection: A New Perspective on What Makes Us Human.* New York: W. W. Norton, 2011.

Shipman, P. *The Invaders: How Humans and Their Dogs Drove Neanderthals to Extinction.* Cambridge, MA: Belknap Press of Harvard University Press, 2015.

Smith, A. An Inquiry into the Nature and Causes of the Wealth of Nations. Hazelton, PA: Pennsylvania State University, 2005. https://eet.pixel-online.org/files/etranslation/original/The%20Wealth%20of%20Nations.pdf.

Smith, B. "General Patterns of Niche Construction and the Management of 'Wild' Plant and Animal Resources by Small-Scale Pre-Industrial Societies." *Philosophical Transactions of the Royal Society* B: Biological Sciences 366 (2011): 836–48.

Smith, B. D., and M. A. Zeder. "The Onset of the Anthropocene." *Anthropocene* 4 (2013): 8–13. doi:10. 1016/j.ancene.2013.05.001.

Smith, E. A., K. Hill, F. W. Marlowe, et al. "Wealth Transmission and Inequality among Hunter-Gatherers." *Current Anthropology* 51 (2010): 19–31.

Snow, C. P. *The Two Cultures.* Cambridge: Cambridge University Press, 2001 (1959).

Snow, D. "Sexual Dimorphism in Upper Paleolithic European Cave Art." *American Antiquity* 78, no. 4 (2013): 746–61.

Sosis, R. 2009. "The Adaptationist-Byproduct Debate on the Evolution of Religion: Five Misunderstandings of the Adaptationist Program." *Journal of Cognition and Culture* 9 (2009): 315–32. doi:10.1163/1567709 09X1251853641411.

Sosis, R., and C. Alcorta. "Signaling, Solidarity, and the Sacred: The Evolution of Religious Behavior." *Evolutionary Anthropology* 12 (2003): 264–74. doi:10.1002/evan.10120.

Spikins, P. *How Compassion Made Us Human: The Evolutionary Origins of Tenderness, Trust, and Morality.* South Yorkshire: Pen & Sword, 2015.

Spikins, P. "Prehistoric Origins: The Compassion of Far Distant Strangers." In *Compassion: Concepts, Research and Applications*, edited by P. Gilbert, 16–30. Abingdon, UK: Taylor and Francis, 2017.

Spikins, P., A. Needham, L. Tilley, and G. E. Hitchens. "Calculated or Caring? Neanderthal Healthcare in Social Context." *World Archaeology* 50, no. 3 (2018): 384–403. https://doi.org/: 10.1080/00438243.2018.1433060.

Spikins, P., H. Rutherford, and A. Needham. "From Homininity to Humanity: Compassion from the Earliest Archaics to Modern Humans." *Time and Mind* 3 (2010): 303–26.

Springer, G., J. Sapp, and A. I. Tauber. "Symbiotic View of Life: We Have Never Been Individuals." *Quarterly Review of Biology* 87, no. 4 (2012): 325–41.

Squire, S. *I Don't: A Contrarian History of Marriage*. New York: Bloomsbury Press, 2008.

Staes, N., C. C. Sherwood, K. Wright, M. de Manuel, E. E. Guevara, T. Marques-Bonet, M. Krützen, et al. "*FOXP2* Variation in Great Ape Populations Offers Insight into the Evolution of Communication Skills." *Scientific Reports* 7, no. 1 (2017): 16866. doi:10.1038/s41598-017-16844-x.

Steffen, W., J. Grinevald, P. Crutzen, and J. McNeil. "The Anthropocene: Conceptual and Historical Perspectives." *Philosophical Transactions of the Royal Society A: Mathematical, Physical and Engineering Sciences* 369 (2011): 842–67.

Sterelny, K. "Artifacts, Symbols, Thoughts." *Biological Theory* 12 (2017): 236–47. DOI 10.1007/s13752-017-0277-3.

Sterelny, K. *The Evolved Apprentice: How Evolution Made Humans Unique*. Cambridge, MA: MIT Press, 2012.

Sterelny, K. "From Hominins to Humans: How Sapiens Became Behaviourally Modern." *Philosophical Transactions of the Royal Society B: Biological Sciences* 366 (2011): 809–22.

Sterelny, K. "A Paleolithic Reciprocation Crisis: Symbols, Signals, and Norms." *Biological Theory* 9, no. 1 (2014): 65–77.

Sterelny, K. *Thought in a Hostile World*. Oxford: Blackwell, 2003.

Sterelny, K., and P. Hiscock. "Symbols, Signals, and the Archaeological Record." *Biological Theory* 9, no. 1 (2014): 1–3.

Stout, D., and T. Chaminade. "Stone Tools, Language and the Brain in Human Evolution." *Philosophical Transactions of the Royal Society B: Biological Sciences* 367 (2012): 75–87.

Stout, D., E. Hecht, N. Khreisheh, B. Bradley, and T. Chaminade. "Cognitive Demands of Lower Paleolithic Toolmaking." *PLoS ONE* 10, no. 4 (2015): e0121804. doi:10.1371/journal.pone.0121804.

Strier, K. B. *Primate Behavioral Ecology*. 5th edition. Abingdon, UK: Routledge/Taylor & Francis, 2016.

Strier, K. B. "What Does Variation in Primate Behavior Mean?" *Yearbook of Physical Anthropology* 162, S63 (2017): 4–14. DOI: 10.1002/ajpa.23143.

Strum, S. "Darwin's Monkey: Why Baboons Can't Become Human." *American Journal of Physical Anthropology* 149, no. S55 (2012): 3–23. https://doi.org/10.1002/ajpa.22158.

Sultan, S. *Organism and Environment: Ecological Development, Niche Construction, and Adaptation.* Oxford: Oxford University Press, 2012.

Sussman, R. W., and C. R. Cloninger. *Origins of Altruism and Cooperation.* New York: Springer, 2011.

Taylor, E. B. *Primitive Culture: Researches into the Development of Mythology, Philosophy, Religion, Language, Art, and Custom.* 2 volumes. 2nd edition. London: John Murray, 1873.

Teleki, G. "They Are Us." In *The Great Ape Project,* edited by P. Cavalieri and P. Singer, 296–302. New York: St. Martin's Griffin, 1993.

Tomasello, M. *The Cultural Origins of Human Cognition.* Cambridge, MA: Harvard University Press, 2009.

Tomasello, M. *A Natural History of Human Thinking.* Cambridge, MA: Harvard University Press, 2014.

Tomasello, M., A. P. Melis, C. Tennie, E. Wyman, and E. Herrmann. "Two Key Steps in the Evolution of Human Cooperation: The Interdependence Hypothesis." *Current Anthropology* 53, no. 6 (2012): 673–92.

Trinkaus, E., and A. P. Buzhilova. "Diversity and Differential Disposal of the Dead at Sunghir." *Antiquity* 92, no. 361 (February 2018): 7–21.

Trut, L., I. Oskina, and A. Kharlamova. "Animal Evolution during Domestication: The Domesticated Fox as a Model." *BioEssays: News and Reviews in Molecular, Cellular and Developmental Biology* 31, no. 3 (2009): 349–60. doi:10.1002/bies.200800070.

Ullah, I. I. T., I. Kuijt, and J. Freemanc. "Toward a Theory of Punctuated Subsistence Change." *PNAS* 112, no. 31 (2015): 9579–84. doi/10.1073/pnas.1503628112.

Ungar, P. S., F. E. Grine, and M. F. Teaford, "Diet in Early *Homo*: A Review of the Evidence and a New Model of Adaptive Versatility." *Annual Review of Anthropology* 35 (2006): 209–28.

Van Anders, S. M. "Beyond Masculinity: Testosterone, Gender/Sex, and Human Social Behavior in a Comparative Context." *Frontiers in Neuroendocrinology* 34, no. 3 (2013): 198–210.

Van Huyssteen, J. W. *Alone in the World? Human Uniqueness in Science and Theology.* Grand Rapids: Eerdmans, 2006.

Van Huyssteen, J. W. "Lecture One: Rediscovering Darwin for Theology—Rethinking Human Personhood." *HTS Teologiese Studies / Theological Studies* 73(3) (2017): a4485. https://doi.org/10.4102/hts.v73i3.4485.

Van Huyssteen, J. W. "Lecture Three: From Empathy to Embodied Faith: Interdisciplinary Perspectives on the Evolution of Rreligion." *HTS Teologiese Studies / Theological Studies* 73, no. 3 (2017): a4488. https://doi.org/10.4102/hts.v73i3.4488.

Van Schaik, C. P. *The Primate Origins of Human Nature.* Oxford: Wiley-Blackwell, 2016.

Von Uexküll, J. *A Foray into the Worlds of Animals and Humans.* Minneapolis: University of Minnesota Press, 2010 (1934).

Wadley, L. "Recognizing Complex Cognition through Innovative Technology in Stone Age and Palaeolithic Sites." *Cambridge Archaeological Journal* 23, no. 2 (June 2013): 163–83.

Wake, D. B., E. A. Hadly, and D. D. Ackerly. "Biogeography, Changing Climates, and Niche Evolution." *PNAS* 106, S2 (2009): 19631–36.

Walker, A., M. R. Zimmerman, and R. E. F. Leakey. "A Possible Case of Hypervitaminosis A in *Homo erectus.*" *Nature* 296 (1982): 248–50.

Waterston, A. "Marriage and Other Arrangements." Open Anthropology 1, no. 1 (2013). http://www.americananthro.org/StayInformed/OAIssueTOC.aspx.

Weiss, K., and A. Buchanan. *The Mermaid's Tale: Four Billion Years of Cooperation in the Making of Living Things.* Cambridge, MA: Harvard University Press, 2009.

Whitehouse, H., P. François, P. E. Savage, T. E. Currie, K. C. Feeney, E. Cioni, R. Purcell, et al. "Complex Societies Precede Moralizing Gods Throughout World History." *Nature* 568 (2019): 226–229.

Whiten, A. "The Scope of Culture in Chimpanzees, Humans and Ancestral Apes." *Philosophical Transactions of the Royal Society B: Biological Sciences* 366 (2011): 997–1007. DOI: 10.1098/rstb.2010.0334.

Whiten, A., F. J. Ayala, M. W. Feldman, and K. N. Laland. "The Extension of Biology through Culture." *PNAS* 114, no. 30 (July 25, 2017): 7775–81.

Whiten, A., R. A. Hinde, C. B. Stringer, and K. N. Laland. *Culture Evolves.* Oxford: Oxford University Press, 2012.

Whiten, A., and E. van de Waal. "Social Learning, Culture and the 'Socio-Cultural Brain' of Human and Non-Human Primates." *Neuroscience and Biobehavioral Reviews* 82 (2017): 58–75. Special issue: "Primate Social Cognition"

Wiessner, P. W. "Embers of Society: Firelight Talk among the Ju/'hoansi Bushmen." *PNAS* 111, no. 39 (2014): 14027–35.

Wiessner, P. W. "The Vines of Complexity: Egalitarian Structures and the Institutionalization of Inequality among the Enga." *Anthropology* 43, no. 2 (2002): 233–69.

Wildman, W. J. *Science and Religious Anthropology.* Farnham, UK: Ashgate. 2009.

Wilkins, J. *Species: A History of the Idea, Species, and Systematics.* Berkeley: University of California Press, 2009.

Wilson, M. L. "Chimpanzees, Warfare and the Invention of Peace." In *War, Peace, and Human Nature: The Convergence of Evolutionary and Cultural Views*, edited by D. P. Fry. 361–88. Oxford: Oxford University Press, 2013.

Wood, B. "Reconstructing Human Evolution: Achievements, Challenges, and Opportunities." *PNAS* 107, no. 2 (2010): 8902–9.

Wood, B., and E. Boyle. "Hominin Taxic Diversity: Fact or Fantasy?" *Yearbook of Physical Anthropology* 159 (2016): S37–S78

Wrangham, R., and R. Carmody. "Human Adaptation to the Control of Fire." *Evolutionary Anthropology: Issues, News, and Reviews* 19, no. 5 (2010): 187–99.

Zeder, M. A. "The Domestication of Animals." *Journal of Anthropological Research* 68 (2002): 161–90.

Zeder, M. A. "Domestication as a Model System for the Extended Evolutionary Synthesis." *Interface Focus* 7 (2017): 20160133.

Zeder, M. A. "The Neolithic Macro-(R)evolution: Macroevolutionary Theory and the Study of Culture Change." *Journal of Archaeological Research* 17 (2009): 1–63. DOI 10.1007/s10814-008-9025-3.

Zeder, M. A. "Why Evolutionary Biology Needs Anthropology: Evaluating Core Assumptions of the Extended Evolutionary Synthesis." *Evolutionary Anthropology: Issues, News, and Reviews* 27, no. 6 (2018): DOI: 10.1002/evan.21747.

Websites

https://www.aeaweb.org/resources/students/what-is-economics.

http://www.americananthro.org/StayInformed/OAIssueTOC.aspx?ItemNumber=2470.

http://www.bbc.com/future/story/20121016-is-language-unique-to-humans.

https://bigthink.com/floating-university/humans-make-language-language-makes-us-human.

https://bizarro.com/2016/09/04/bush-voyeurs/.

https://www.britannica.com/topic/economics#ref236757.

https://www.dalailama.com/messages/compassion-and-human-values/compassion.

https://eet.pixelonline.org/files/etranslation/original/The%20Wealth%20of%20Nations.pdf

https://evolution.berkeley.edu/evolibrary/article/evo_25.

https://www.focus-economics.com/blog/the-largest-economies-in-the-world.

https://www.giffordlectures.org/file/terry-eagleton-god-debate.

https://www.giffordlectures.org/lectures/making-representations-religious-faith-and-habits-language.

https://www.giffordlectures.org/lectures/man-his-nature.

https://www.google.com/search?q=beleif+dictionary&rlz=1C1GGRV_en

US751US751&oq=beleif+dictionary&aqs=chrome..69i57j0l5.3479j0j4&
sourceid=chrome&ie=UTF-8.

https://healthinequality.org/.

http://johnhawks.net/weblog/topics/news/finlayson-braided-stream-2013.html.

https://kevishere.com/2014/06/10/developmental-plasticity-and-the-hard
-wired-problem/.

https://www.merriam-webster.com/dictionary/economy.

http://www.mcescher.com/.

http://money.cnn.com/2015/07/08/news/economy/global-low-income/index
.html.

http://www.monkeyforestubud.com/concept-of-monkey-forest/the-tri
-hita-karana/.

https://www.nature.com/news/diversity-1.15913.

https://www.nature.com/subjects/behavioural-ecology.

https://www.nytimes.com/2018/04/30/opinion/karl-marx-at-200-influence
.html.

https://www.oxfam.org/en/even-it/5-shocking-facts-about-extreme-global
-inequality-and-how-even-it-davos.

http://www.pewforum.org/2012/12/18/global-religious-landscape-exec/.

http://www.pewresearch.org/topics/discrimination-and-prejudice/.

http://www.pewresearch.org/fact-tank/2016/07/01/racial-gender-wage-gaps
-persist-in-u-s-despite-some-progress/.

http://www.pewresearch.org/fact-tank/2017/12/08/global-snapshot-sex
-marriage/.

http://www.pewresearch.org/topics/immigration-attitudes/2018/.

http://www.pewresearch.org/topics/income-inequality/.

http://www.pewsocialtrends.org/.

https://plato.stanford.edu/entries/economics/#1.

https://plato.stanford.edu/entries/marx/.

https://plato.stanford.edu/entries/love/.

https://plato.stanford.edu/entries/smith-moral-political/.

https://www.psychologytoday.com/us/blog/animal-emotions/201808
/make-no-mistake-orca-mom-j-35-and-pod-mates-are-grieving.

https://www.theatlantic.com/technology/archive/2012/11/noam-chomsky
-on-where-artificial-intelligence-went-wrong/261637/.

https://www.theguardian.com/science/punctuated-equilibrium/2010/oct/20/3.

https://www.theguardian.com/us-news/2018/jun/06/growing-up-black
-in-america-racism-education.

http://www.understandingrace.com.

http://wallacefund.info/content/1858-darwin-wallace-paper.

https://en.wikipedia.org/wiki/Belief.

https://en.wikipedia.org/wiki/Crittercam.

https://en.wikipedia.org/wiki/Market_economy.

http://www.worldometers.info/world-population/.

Index